Posture, locomotion, and paleoecology of pterosaurs

by
Sankar Chatterjee
Museum of Texas Tech University
Box 43191
Lubbock, Texas 79409-3191
USA

and

R.J. Templin
2212 Aster Street
Ottawa, Ontario K1H 6R6
Canada

THE GEOLOGICAL SOCIETY OF AMERICA

Special Paper 376

3300 Penrose Place, P.O. Box 9140 ▪ Boulder, Colorado 80301-9140 USA

2004

Published by The Geological Society of America, Inc.
3300 Penrose Place, P.O. Box 9140, Boulder, Colorado 80301
www.geosociety.org

Printed in U.S.A.

GSA Books Science Editor: Abhijit Basu

Library of Congress Cataloging-in-Publication Data

Chatterjee, Sankar.
Posture, locomotion, and paleoecology of pterosaurs / by Sankar Chatterjee, R.J. Templin.
p. cm. — (Special paper / Geological Society of America ; 376)
Includes bibliographic references.
ISBN 0-8137-2376-0 (pbk)
1. Dinosaurs--Flight. 2. Pterosauria. 3. Dinosaurs--Evolution. 4. Birds--Evolution. 5. Paleontology--Mesozoic. I. Templin, R. J. II. Title. III. Special papers (Geological Society of America) ; 376.

QE861.6.F45C48 2004
567.918—dc22

2003064301

Cover: A flock of Anhanguera flying in V-formation like modern geese to save energy. Painting by Michael W. Nickell.

10 9 8 7 6 5 4 3 2 1

Contents

Geological Society of America
Special Paper 376
2004

Posture, locomotion, and paleoecology of pterosaurs

Sankar Chatterjee*
Museum of Texas Tech University, Box 43191, Lubbock, Texas 79409-3191, USA

R.J. Templin*
2212 Aster Street, Ottawa, Ontario, Canada K1H 6R6

ABSTRACT

Despite being studied for over 200 years, the posture, locomotion, paleobiology, and phylogeny of pterosaurs are still poorly understood due to lack of well-preserved, three-dimensional specimens. In this paper we investigate the posture, locomotion, and paleoecology of pterosaurs based on anatomy and biomechanics: how they walked, how they flew, and how they lived. We want to understand how evolution has adjusted their skeletal structures and movements to maximize performance. The limb joints of an exquisite skeleton of the Cretaceous pterodactyl *Anhanguera piscator* are analyzed to estimate the range of movement during terrestrial and aerial locomotion. On land, pterosaurs were quadrupedal knuckle walkers with laterally directed digitigrade manus and forwardly directed plantigrade pes. From this position, pterosaurs could stand on their rear legs and run bipedally in an upright posture for a short distance during takeoff and landing.

Pterosaurs evolved two basic wing planforms over time: the basal "rhamphorhynchoids" had broad wings in bat-like fashion where the patagium was attached to the ankle; in pterodactyloids, the wings became narrow and the patagium was anchored near the knee joint. The wingtips appear to have been more rounded to avoid stalling. The actinofibrils in the membrane would confer some stiffness to the wing to maintain a flatter camber, preventing it from billowing and tearing during flight. They would also facilitate the folding of the wing when not in use.

The flight performance of pterosaurs is investigated using ten genera in a wide size spectrum during their 160 million years of evolution, where the body mass ranges from 0.015 kg to 70 kg and the wingspan from 0.4 m to 10.4 m. Thus the largest pterosaur in our study weighs about 4700 times more than the smallest species, and the longest wingspan is 25 times the shortest. Helicopter momentum stream tube theory has been adapted to estimate the scaling of aerial locomotion of pterosaurs and to minimize the complexities of animal physiology. The aerodynamic data were calculated using the two computer programs ANFLTPWR (animal flight power) and ANFLTSIM (animal flight simulation). Pterosaur wings were long and narrow, similar to those of seabirds, with high aspect ratios. However, they had relatively low wing loadings and low cruising speeds compared to seabirds with similar masses. Gliding performance, deduced from the polar curves, indicates that smaller pterosaurs such as *Eudimorphodon*, *Pterodactylus*, *Rhamphorhynchus*, and *Dorygnathus* had lower gliding airspeeds, with a gliding angle close to 4°. The giant Cretaceous pterodactyloids such as *Pteranodon* and *Quetzalcoatlus* were excellent soarers comparable to the albatross, human-powered planes, and sailplanes, with a gliding angle between 1° and 2°. The cruising speed for best gliding depends on size, increasing proportionally to mass and wing loading,

*E-mails: sankar.chatterjee@ttu.edu, jtemplin@magma.ca

Chatterjee, S., and Templin, R.J., 2004, Posture, locomotion, and paleoecology of pterosaurs: Boulder, Colorado, Geological Society of America Special Paper 376, p. 1–64. For permission to copy, contact editing@geosociety.org.

from as low as 4 m/s for *Eudimorphodon* to 16 m/s for *Quetzalcoatlus*. The power curves, displaying maximum and minimum level flight speeds, show three different styles of flight. In the four smaller genera (Mass < 0.3 kg) such as *Eudimorphodon*, *Pterodactylus*, *Rhamphorhynchus*, and *Dorygnathus*, the available aerobic power (Pa) exceeds the required power at zero speed, and they were evidently capable of hovering flight. *Tapejara*, *Nyctosaurus*, *Dsungaripterus*, *Anhanguera*, and *Pteranodon* appear to be capable of steady level flight at aerobic power, but within a limited speed range. The sustained power output of giant pterodactyloids such as *Quetzalcoatlus* was not enough for continuous level flapping flight; however, they could improve their flying performance if they flew in formation. Apparently, extended flight for large pterodactyloids was by soaring; they flapped normally when taking off or landing. Takeoff from the ground was initiated by running and hopping. Although small pterosaurs apparently had sufficient available power for running takeoff from the ground or from the water, larger pterodactyloids such as *Pteranodon* and *Quetzalcoatlus* were limited in their takeoff capabilities and were unable to take off with maximum aerobic power. They needed short bursts of anaerobic power to take off from the ground with a headwind of 5 m/s. For *Quetzalcoatlus*, a takeoff from a 10° downward slope would be helpful especially when it ventured inland. As their running speed increased, low-amplitude flapping was used to accelerate to take off; pterosaurs leaped into the air and flapped their wings for flight. The long axis rotation of the humerus in the upstroke position would have been useful during takeoff and landing.

The function of the cranial crest may have been linked to thermoregulation, sexual display, and species recognition. The large head of pterodactyloids was probably downturned during flight and was used as a steering device for turning the body. The ecology of pterosaurs was similar to those of modern seabirds, spending much time in coastal areas for feeding. Small and medium-size pterosaurs probably foraged by plunge diving like modern pelicans. Large pterodactyloids were probably active waders or surface riders during feeding, using their feet to propel while folding their wings sidewise. Arising as small animals in the Triassic, pterosaurs exhibit long-term phyletic trends toward increasing body size during the Cretaceous, but the trend is erratic. They became extinct at the end of the Cretaceous along with dinosaurs and other organisms when multiple asteroids crashed into the Earth, accompanied by the spectacular Deccan volcanism that had devastating effects on the ecology.

Keywords: pterosaurs, Mesozoic reptiles, terrestrial locomotion, flight performance, wing design.

INTRODUCTION

Among tetrapods, only birds, bats, and pterosaurs evolved independently powered flight by wing flapping. Each group has acquired different styles of flight because of different body plans. Yet, they all used the same aerodynamic principles. But the image of pterosaurs, the prehistoric dragons amidst the clouds, captures our fancy and imagination. Pterosaurs are wonderfully enigmatic archosaurs that display an incredible parade of designs and sizes. They first appeared in the fossil record during the Late Triassic, diversified into an extraordinary variety of forms and sizes during the Jurassic and Cretaceous periods, and dominated the Mesozoic sky for 160 million years. At the same time, however, their wing morphology remained fairly conservative. Early in history, pterosaurs invented a successful style of flying that they refined for the remainder of the Mesozoic to become the largest aerial animals that ever lived (Langston, 1981). They radiated globally as the continents rifted and drifted apart, and became extinct at the end of the Cretaceous along with the dinosaurs and many other animals. Their fossils have been found on every continent, including Antarctica.

Pterosaurs have a unique wing anatomy unlike that of bats and birds. The wing membrane is supported by the forelimb and one hyper-elongated fourth finger; the latter comprises at least half of the total wing length. Note that pterosaur digits one, two, and three maintain their primitive archosaurian proportions and sustain claws; the fourth is unusually long, whereas the fifth digit is lost. Although Collini, the keeper of the Mannheim Natural History Collection, described the first pterosaur fossil in 1784 from the Late Jurassic Solnhofen Limestone of Eichstätt, Germany, 40 years before the first discovery of the dinosaurs, many aspects of pterosaur biology still remain unknown. Collini (1784) assigned this enigmatic animal to an unknown marine sea creature. It was Cuvier (1801), the famous

French anatomist and father of vertebrate paleontology, who correctly understood the reptilian affinity and flying ability of the Eichstätt specimen. Cuvier's felicitous name for the creature was *Pterodactylus*—the wing finger—after its elongated fourth digit. Since then, thousands of pterosaur specimens have been discovered from all over the world, but their anatomy and ecology remain enigmatic and controversial. It could be argued that the greatest contributions to pterosaur anatomy and flight performance have surfaced within the past three decades. Still, after much debate, many workers generally do not agree on even the simplest aspect of pterosaur paleobiology and design. For example, did the wing membrane caudally attach to the body or the hindlimbs? Was the hindlimb held erect or was it semi-sprawling? Were the pterosaurs quadrupedal or bipedal? Did they glide exclusively or flap their wings? Could they take off from the ground? Debate continues regarding how pterosaurs moved when on the ground and how maneuverable they were in the air. These and other questions pertaining to their terrestrial and aerial locomotion are explored below.

Considering the large number of taxa (~120 species) that evolved throughout the Mesozoic, pterosaurs are all remarkably similar in their general body form, but differ greatly in size. Their most variable parts seem to be their two ends—skulls and tails. The skulls are proportionately large and elongated, with large orbits, and external nares are set far back from the tip of the beak. They are lightly built, pneumatic and akinetic. Some species of pterosaurs sported peculiar crests on both of their upper and lower jaws, others had one on the skull; some entirely lacked any crests. Like Mesozoic birds, pterosaurs show two morphotypes: long-tailed and short-tailed forms (Chatterjee, 1997).

Based on skull and tail morphology, pterosaurs have been traditionally divided into two groups—"rhamphorhynchoids" and pterodactyloids (Romer, 1956; Wellnhofer, 1991a). The "rhamphorhynchoids" evolved first and are currently considered a paraphyletic group (Padian, 1983; Unwin, 1995). Many of the "rhamphorhynchoids" were small, about the size of pigeons or gulls with oversized wings. They are the basal group of early pterosaurs with a short skull and neck, a long tail with elongate zygapophyses and chevrons, and a long fifth toe. The bony tails can be very long and stiff with a rudder-like vertical vane on the tip, which perhaps functioned as a dynamic stabilizer in flight. In this group, the external naris is separated from the antorbital fenestra by a bony bridge as in other archosaurs. The "rhamphorhynchoids" first appeared in the Late Triassic (Wild, 1978) and continued through the end of the Jurassic. The location of the occipital condyle on the longitudinal axis of the skull indicates that they carried their heads horizontally in a rather extended position (Wellnhofer, 1975).

During the Late Jurassic, the pterodactyloids gradually replaced the "rhamphorhynchoids" and flourished throughout the Cretaceous. The pterodactyloids are a monophyletic clade and more advanced than the "rhamphorhynchoids" (Padian, 1983; Unwin, 1995; Unwin and Lu, 1997; Bennett, 1994). They have a long skull and neck, highly inclined quadrate, and a tendency towards reduction in the number and size of teeth. In this group, the external naris becomes confluent with the antorbital fenestra. The occipital condyle in this group lies on the lower side of the skull suggesting that the head was carried at a distinct angle to the neck (Wellnhofer, 1991a). The fifth toe becomes relatively short due to the loss of digits. Pterodactyloids more or less dispensed with the tail; the tailless control is more efficient aerodynamically even if more difficult to operate (Pennycuick, 1972). The truncated tail in this group led to instability in flight, as in bats and birds, in favor of greater maneuverability (Chatterjee, 1997). It would reduce the stalling speed of large pterodactyloids thus making it easier to land. Pterodactyloids were more variable in size; some Jurassic forms were as small as a crow, but Late Cretaceous members such as *Pteranodon* and *Quetzalcoatlus* became the largest flying animals with a wingspan ranging from 7 to 10.4 m. *Quetzalcoatlus* was not only the largest known animal that ever flew (Langston, 1981), but it also pushed the boundary of animal flight.

Ever since Cuvier (1801) recognized the flying ability of pterosaurs from the wing structure, their flight performance has captured enormous interest for the last two centuries. Because pterosaurs are extinct, ideas about their flying style are largely speculative; most interpretations are inferred from morphologic correlates and comparative study of modern analogs such as bats and birds. How did pterosaurs fly? Did they glide exclusively or flap their wings or even hover? In this paper we analyze various modes of flight performances of pterosaurs using a wide range of genera.

The aerodynamic analysis of pterosaurs is often marred by their two-dimensional preservation. Pterosaur bones are extremely light, hollow, pneumatic, and thin-walled—perfect for flight but not for preservation. Acid preparation of two beautiful skulls at the Museum of Texas Tech University, *Rhamphorhynchus* (CM 11431) and *Anhanguera* (AMNH 25555), clearly indicates how pterosaur skulls are made of paper-thin sheets of bone, fused together and stiffened inside with thin struts and reticulating ridges (Chatterjee, 1992; Witmer et al., 2003). The skeletons of pterosaurs are generally preserved as crushed and fragmented sheets, affording little information on joint movements. This deficiency of preservation has been remedied in recent years with the discovery of a series of excellent specimens from the Early Cretaceous Santana Formation of Brazil. They are exquisitely preserved, intact and uncrushed within the undisturbed limestone sequence at the bottom of the lagoon. Recent acquisition of three superb, three-dimensional casts of pterosaur skeletons for the upcoming dinosaur exhibit at the Museum of Texas Tech University prompted this study. They provide important data for the terrestrial and aerodynamic performance of pterosaurs.

Flight imposes limitations on size and body weight. Unlike bats, both birds and pterosaurs exhibit the largest range of sizes among flying vertebrates. The mass of the largest bat, such as the fruit bat *Pteropus* is only 1.6 kg with a wingspan up to 1.7 m (McFarland et al., 1979). By comparison, the largest known flying bird, *Argentavis magnificens* from the Late Miocene of Argentina, had a wingspan of 7.5 m and weighed as much as 62 kg (Campbell and Tonni, 1983; Viscain and Farina, 1999). Among living soaring birds, the albatross has a wingspan of 3.4 m and mass close to 8.5 kg. The largest known pterosaurs, *Quetzalcoatlus northorpi* (Lawson, 1975; Langston, 1981; Brower and Veinus, 1981; Atanassov and Strauss, 2002), had a wingspan of 10.4 m and a mass of about 70 kg. In comparison, the Boeing 747 has a wingspan of 60 m and a mass of 360,000 kg (Tennekes, 1996), whereas the six-engined Antonov An-225 Myira has a wingspan over 88 m and weighs more than 600,000 kg (Paul, 2002). *Quetzalcoatlus* weighs about 4700 times more than the smallest species of pterosaurs. In contrast, Antonov An-225 is about 2200 times heavier than the Wright 1903 *Flyer*.

Several morphological features such as the wing geometry, wing support, wing movement, structural fibers of the patagium, and the animal's stance are crucial in understanding their flight performance, their ability to take off and land, and their ecology. We have addressed these issues in the beginning to show the similarities and differences between pterosaurs and their living analogs, bats and birds, before undertaking any aerodynamic analysis. Our aim is to search for a uniform set of aerodynamic constraints across this wide spectrum. Many workers have interpreted pterosaurs as capable only of gliding (Kripp, 1943; Bramwell and Whitfield, 1974; Stein, 1975; Brower, 1983). The flapping performance of such large pterosaurs has not been investigated adequately except for a brief analysis by Stein (1975) and Brower (1980, 1983). We found that small pterosaurs (Mass < 0.3 kg) were capable of hovering flight. The purpose of this work is to assess the hovering, flapping, gliding, and soaring performances of pterosaurs using modern aerodynamic techniques. The study shows that pterosaurs pioneered many intricate flight techniques that were subsequently adopted by bats and birds.

MATERIALS AND METHODS

Terrestrial and aerodynamic data were derived for ten well-known species of pterosaurs from the United States, Germany, China and Brazil. Among these ten species of pterosaurs, three species—*Eudimorphodon ranzii* (Wild, 1978), *Rhamphorhynchus muensteri* (Wellnhofer, 1975), and *Dorygnathus banthensis* (Padian and Wild, 1992)—belong to the "Rhamphorhynchoidea." The other seven species—*Pterodactylus antiquus* (Wellnhofer, 1970), *Tapejara wellnhoferi* (Kellner, 1989, 1995), *Nyctosaurus gracilis* (Brower and Veinus, 1981), *Dsungaripterus weii* (Young, 1973), *Anhanguera piscator* (Kellner and Tomida, 2000), *Pteranodon longiceps* (= *P. ingens*, sensu Bennett, 1994), and *Quetzalcoatlus northropi* (Langston, 1981)—belong to the Pterodactyloidea. A tentative phylogeny of these ten species of pterosaurs is shown in Figure 1 based on recent cladistic analysis. They represent a wide spectrum of size, phylogeny, and geological range over the broad evolutionary history of pterosaurs (Tables 1, 2). Some of the data are taken from published literature, others from direct measurements. Three excellent three-dimensional skeletons (casts)—*Tapejara* (TTU P10362; Fig. 2) and *Anhanguera* (TTU P10363, Fig. 3) and a skull (cast) of *Thalassodromeus* (TTU P10397; Fig. 36) from the Early Cretaceous Santana Formation of Brazil (vendor: Museu National, Rio de Janeiro, Brazil); and *Quetzalcoatlus* (TTU P10390; Fig. 4) from the Late Cretaceous Javelina Formation in Big Bend National Park of Texas (vendor: Natural History Exhibit Hall, Livingston, Montana) were purchased for the upcoming dinosaur exhibit at The Museum of Texas Tech University. In addition, a nearly complete skeleton of *Rhamphorhynchus* (CM 11431; Fig. 5) from the collection of the Carnegie Museum of Natural History has been acid prepared at Texas Tech University and is now entirely free from matrix. Direct measurements of wingspan and other skeletal dimensions were

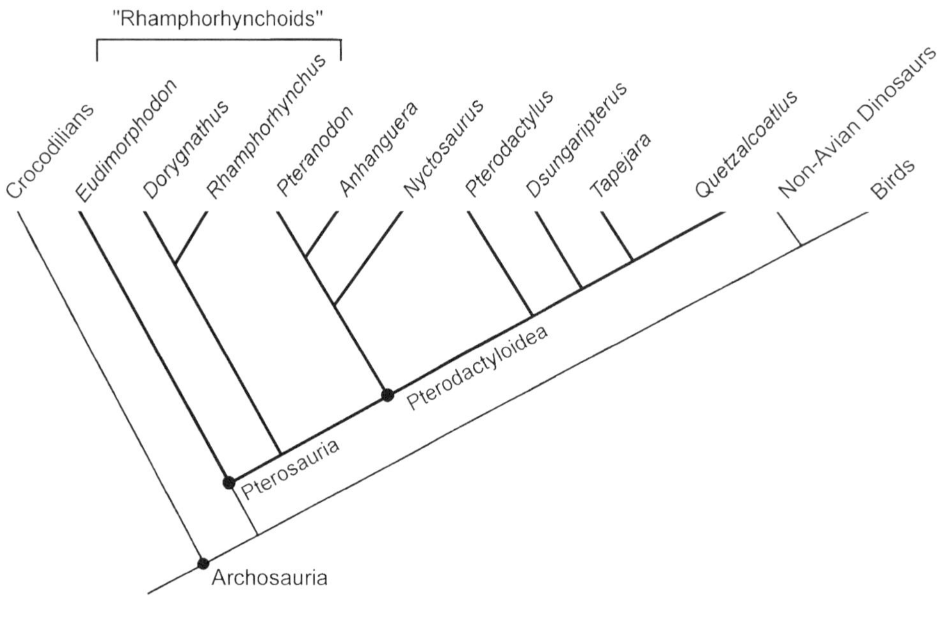

Figure 1. Phylogeny of selected genera of pterosaurs used in this study (simplified from Unwin and Lu, 1997).

TABLE 1. REFERRED SPECIMENS OF 10 SPECIES OF PTEROSAURS

Species	Specimen	Reference	Country	Formation	Age
1. *Eudimorphodon ranzii*	MCSNB 2888	Wild, 1978	Italy	Calcare di Zorzino	U. Triassic
2. *Pterodactylus antiquus*	BSP AS 1739	Wellnhofer, 1970	Germany	Solnhofen	U. Jurassic
3. *Rhamphorhynchus muensteri*	SMF R4128	Wellnhofer, 1975	Germany	Solnhofen	U. Jurassic
4. *Dorygnathus banthensis*	SMNS 51827	Hazelhurst & Rayner (1992a)	Germany	Solnhofen	U. Jurassic
5. *Tapejara wellnhoferi*	MN CD-R 109 (TTU P10362)	Kellner (1989, 1995)	Brazil	Santana	L. Cretaceous
6. *Nyctosaurus gracilis*	YPM 1178	Brower & Veinus (1981)	USA	Niobrara	U. Cretaceous
7. *Dsungaripterus weii*	IVPP 64041-3	Brower & Veinus (1981)	China	Tugulu	L. Cretaceous
8. *Anhanguera piscator*	NSM-PV 19892 (TTU P10363)	Kellner & Tomida (2000)	Brazil	Santana	L. Cretaceous
9. *Pteranodon longiceps*	YPM 1175	Bramwell & Whitfield (1974)	USA	Niobrara	U. Cretaceous
10. *Quetzalcoatlus northropi*	(TTU P10390)	Brower & Veinus (1981)	USA	Javelina	U. Cretaceous

TABLE 2. FORELIMB/HINDLIMB INDICES OF 10 SPECIES OF PTEROSAURS

Species	Humerus	Radius	Mc IV	Femur	Tibia	Limb Index
1. *Eudimorphodon ranzii*	26	35	9.8	19	25	160.9
2. *Pterodactylus antiquus*	31.5	47	35	34.7	48.3	136.7
3. *Rhamphorhynchus muensteri*	38	66	21	33	49	152.4
4. *Dorygnathus banthensis*	56	90	28	45	58	168.9
5. *Tapejara wellnhoferi*	71.6	100.7	96	88.2	117.7	130.3
6. *Nyctosaurus gracilis*	106.1	165	266.4	101	162.7	203.5
7. *Dsungaripterus weii*	125	300	300	221	270	147.7
8. *Anhanguera piscator*	252	384	256	232	282	173.5
9. *Pteranodon longiceps*	285.5	431	665.4	269	372.5	215.4
10. *Quetzalcoatlus northropi*	531	670	1338	899	1340	113.4

Note: Forelimb/hindlimb index = [(humerus + radius + McIV) × 100/(femur + tibia)].

made from these specimens. Many of the skeletal joints of the cast specimens, especially *Anhanguera*, were manipulated to understand the range of motion in the wing and hindlimb elements. Some aerodynamic data were generated by using two computer programs: ANFLTPWR (animal flight power) and ANFLTSIM (animal flight simulation) (see Templin, 2000).

Institutional Abbreviations

BSP—Bayerische Staatssammlung für Paläontologie und historiche Geologie in Munich, Germany

CM—Carnegie Museum of Natural History, Pittsburgh, Pennsylvania

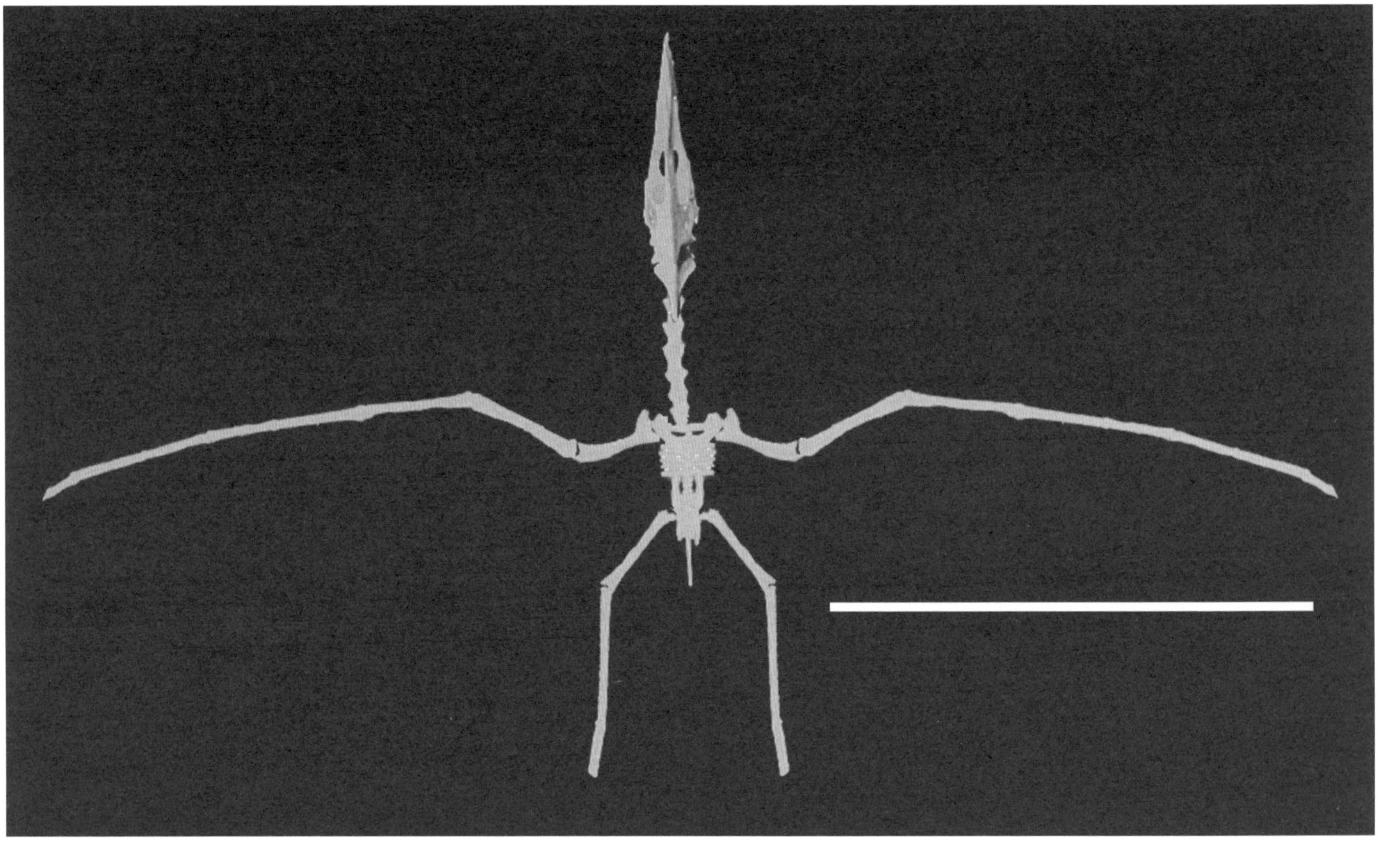

Figure 2. Ventral view of the cast skeleton of *Tapejara wellnhoferi* (TTU P10362); the wingspan is 1.35 m; scale bar = 0.5 m.

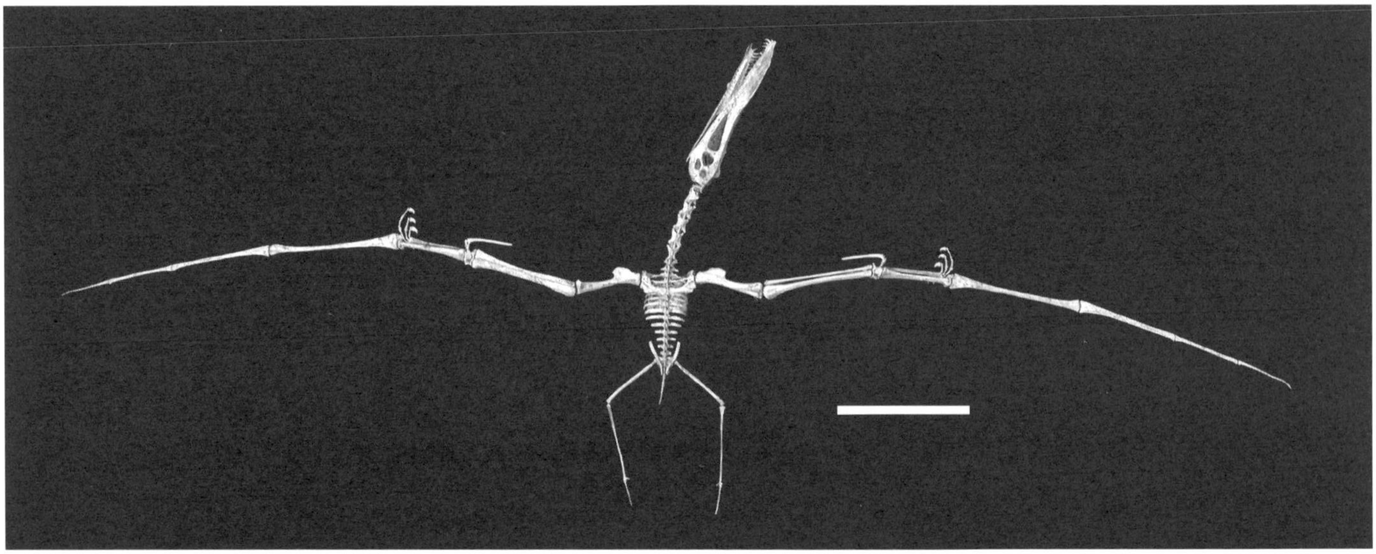

Figure 3. Dorsal view the cast skeleton of *Anhanguera piscator* (TTU P10363); the wingspan is 4.69 m; scale bar = 0.5 m.

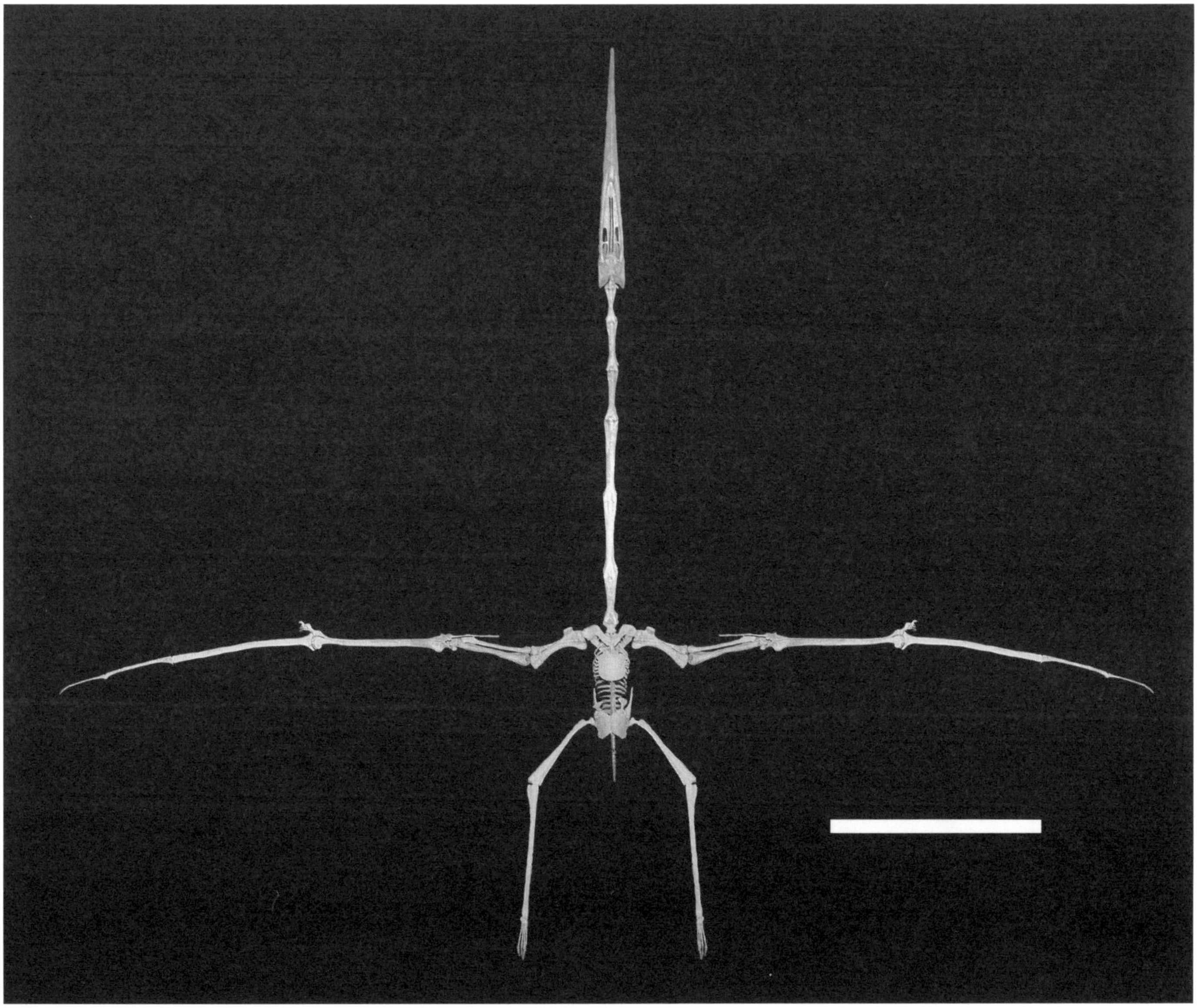

Figure 4. Ventral view of the cast skeleton of *Quetzalcoatlus northropi* (TTU P10390); the wingspan is 10.39 m; scale bar = 2 m.

IVPP—Institute of Vertebrate Paleontology and Palaeoanthropology in Beijing, China

MN—Museu National, Rio de Janeiro, Brazil

MSCNB—Museo Civico di Scienze Naturali in Bergamo, Italy

NHMW—Naturhistorische Museum, Vienna, Austria

NSM—National Science Museum, Tokyo, Japan

SMF—Senckenberg Museum in Frankfurt, Germany

SMNS—Staatliches Museum für Naturkunde in Stuttgart, Germany

TTU—Museum of Texas Tech University, Lubbock, Texas, USA

YPM—Yale Peabody Museum, New Haven, USA

TERRESTRIAL LOCOMOTION

Stance and Gait

The posture and mode of terrestrial locomotion of pterosaurs are still highly controversial. An extinct animal's posture and locomotion cannot be observed directly. They may be inferred from limb mechanics and joint surfaces, as well as from fossil trackways. The debate on pterosaur posture has been polarized into two opposing views: whether pterosaurs walked quadrupedally like bats or bipedally like theropods and birds. Our work suggests that both models are partially correct. Pterosaurs adopted both modes of locomotion: quadrupedal

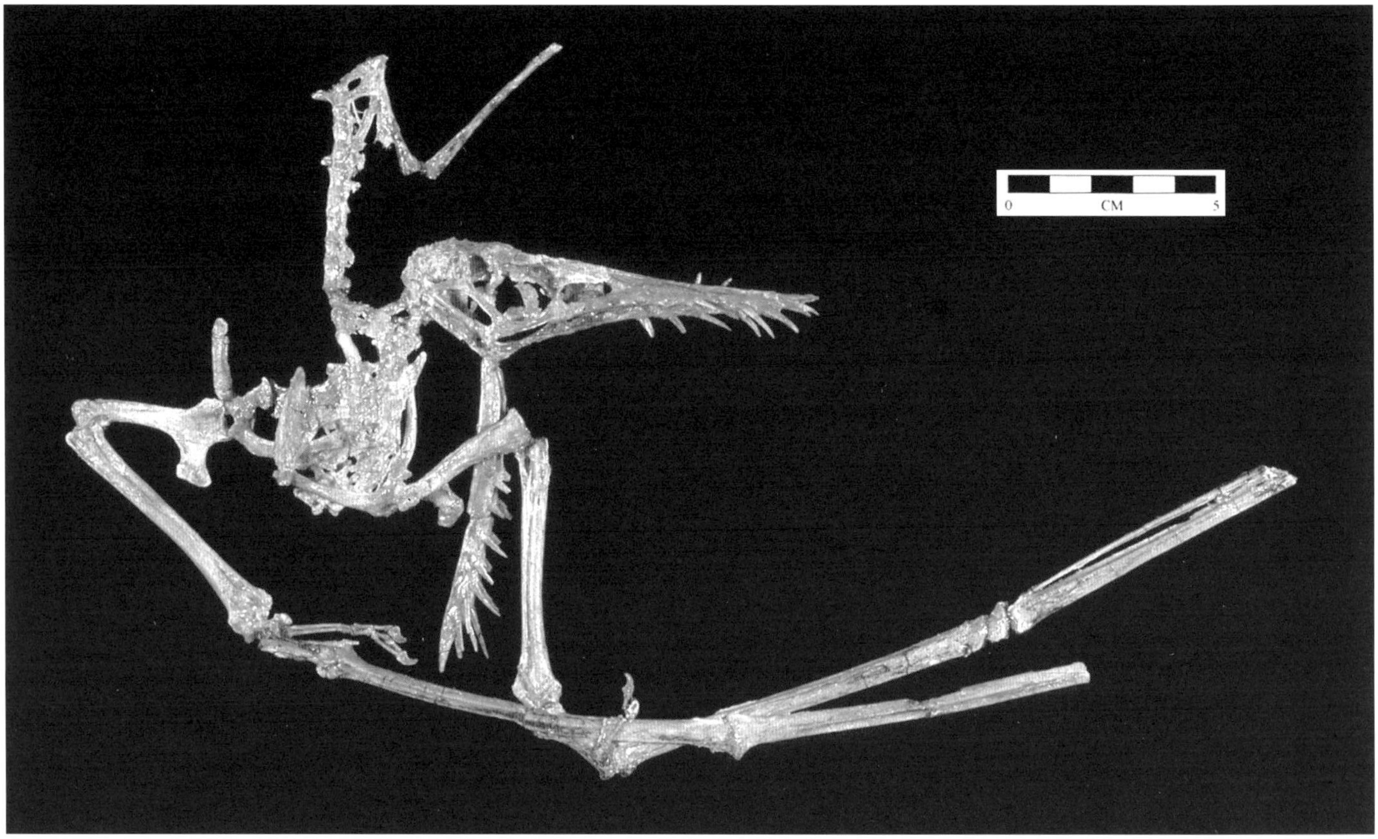

Figure 5. Lateral view of the skeleton of *Rhamphorhynchus muensteri* (CM 11431); the estimated wingspan is 0.86 m; scale bar = 5 cm.

during slow walking, but bipedal for a short burst during take-off and landing.

Traditional reconstructions of pterosaurs show them as clumsy and bat-like, walking on the ground on all fours, with a sprawling gait and a plantigrade pes (Abel, 1925). An alternative view suggests that pterosaurs were obligatory bipedal cursors like theropods and birds where the gait was parasagittal and the stance digitigrade (Seeley, 1870; Padian, 1983; Padian and Rayner, 1993). Although modern analogs are useful to infer the biology of extinct animals, we have to remember that pterosaur anatomy is unique and quite different from that of birds and bats. The theropod model has been challenged recently on the basis of the structure of the hip joint, hindlimb morphology, pterosaur footprints, new fossil evidence, and computer simulation (Wellnhofer, 1988, 1991a, 1991b; Unwin, 1987, 1988a, 1989; Unwin and Bakhurina, 1994; Unwin and Henderson, 1999; Bennett, 1997a, 1997b; Lockley et al., 1995; Clark et al., 1998). The new view favors the traditional interpretation that pterosaurs were quadrupedal walkers with a flat-footed stance.

Although the quadrupedal posture has gained currency in recent years, there is a disagreement over whether the gait was sprawling with a subhorizontal vertebral column (Wellnhofer, 1988, semi-erect (Lockley et al., 1995), or fully erect with a steeply oriented vertebral column (Unwin and Henderson, 1999; Henderson and Unwin, 1999).

Given the differing interpretations of pterosaur terrestrial locomotion, we examine here the range of motion permitted at the hip, knee, and ankle joints from a three-dimensional cast of *Anhanguera*. Both girdles were kept articulated with the vertebral column to maintain the orientation, but the limb elements were dismantled. All angles of wing joints were measured relative to a set of three axes mutually perpendicular: fore and aft (sagittal), lateral (transverse), and vertical (*sensu* Bramwell and Whitfield, 1974). A three-axis Cartesian coordinate reference system was used for each girdle (Fig. 6). Each segment of the joint was manipulated without undue force and the positions were photographed with a digital camera. The joint angles were measured in reference to these three axes with a protractor as well from the photographs. Using this morphometric and kinematic information, the locomotion of *Anhanguera* was studied graphically using simple computer animation and drawing programs (such as iMovie®, Photoshop®, the Graphite 2 mouse) to assess bipedal and quadrupedal locomotion (McQuilkin et al., 2002). They facilitate the selection of a dynamic pose by allowing a frame to be withdrawn from sequence and used as a blueprint, and they can help test the accuracy of a biomechanical

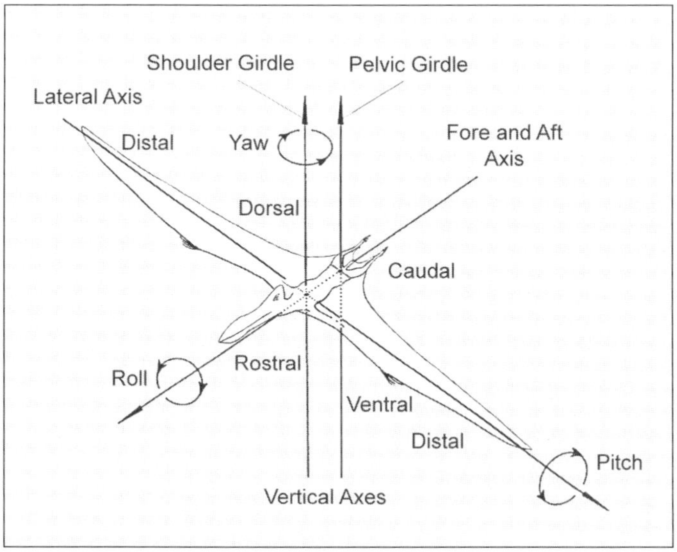

Figure 6. Three axes of orientation of pterosaurs, fore and aft, lateral and vertical (modified from Bramwell and Whitfield, 1974). An additional vertical axis through the pelvic girdle is shown. All the angles of wing and hip joints were measured relative to these axes.

hypothesis. This analysis supports the current view that pterosaurs had a near-erect quadrupedal posture (Wellnhofer, 1988), with long arms and short legs, with digitigrade manus directed laterally but plantigrade pes pointing forward (Bennett 1997a, 1997b). The spine was inclined 45° or more and the upper part of their body was supported on the knuckles and flat fingers. From this position, pterosaurs could stand up on their legs for bipedal running during takeoff.

Forelimb Joints

The movement of the forelimbs during terrestrial locomotion was quite different from that in flight. During normal walking, the elbow would be directed backward and the humerus could be adducted closer to the body, about 25° in relation to the vertical axis. The humerus circumscribed a limited arc of movement ranging from 15° to 45° in relation to the fore and aft axis. Henderson and Unwin (1999) showed in his animation that the forearm and metacarpus moved as a rigid unit during locomotion, keeping the wrist joint fully open at 180°. However, the elbow and wrist joints were automatically coupled in pterosaurs as in birds (Hankin and Watson, 1914; Bramwell and Whitfield, 1974; Wellnhofer, 1985, 1991a). As a result, as soon as the elbow was flexed, the wrist would be flexed too. It could not remain fully open during terrestrial locomotion. The flexion and extension of the manus was restricted to the knuckle joint, between the metacarpal IV and the first phalanx. The manus tracks are oriented laterally indicating lateral torsion of the hand (Lockley et al., 1995). However, the track widths of the manus and pes are narrow, indicating that both forelimbs and hindlimbs were held close to the body in a near-vertical quadrupedal gait (Bennett, 1997a). The flight digit was folded subparallel to the metacarpal four and tucked up behind the pelvis (Fig. 7).

Hip Joint

All pterosaurs have a uniform pelvic morphology, primitively designed like that of basal archosaurs, and fundamentally distinct from the fully erect, bipedal theropods. In pterosaurs, the acetabulum is imperforate, the supracetabular crest is atrophied, the pubo-ischiadic plate is short and broad, and there is an extra element, the pre-pubis. The ilium has a very shallow dorsal blade

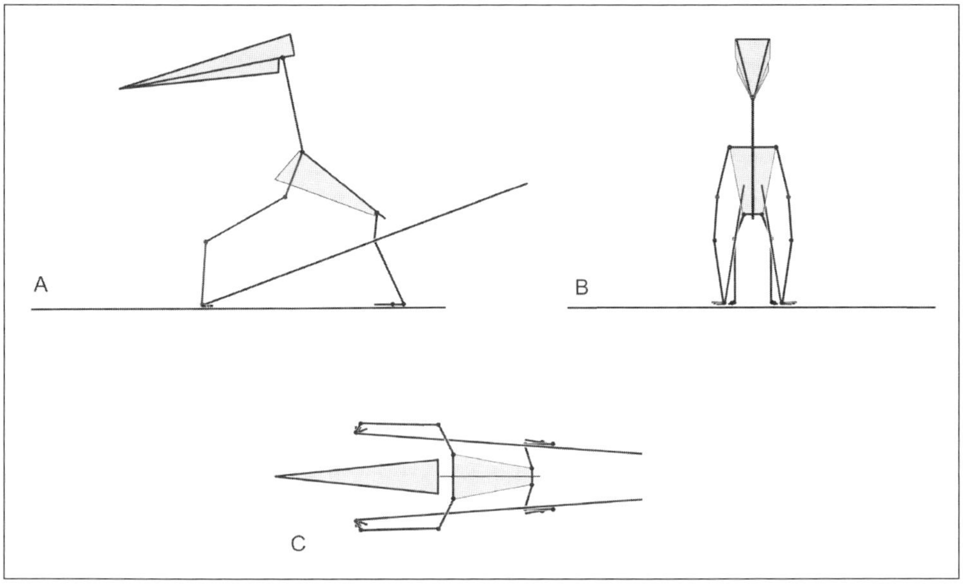

Figure 7. Stick diagram showing the quadrupedal pose of *Anhanguera*. A: Left lateral view. B: Cranial view. C: Dorsal view.

and its preacetabular process is very long and narrow for attachment of protractor muscles. The two halves of the pelvis are fused along the ischial symphysis in all pterosaurs (contra, Wellnhofer, 1988) making the hindlimb closer to the body.

Padian (1979, 1983, 1985, 1988) argued that the hindlimb of pterosaurs was functionally analogous to that of bipedal theropods. However, anatomical comparison of hip joints suggests that theropods and pterosaurs acquired different locomotory styles. In theropods, the femoral component is cylindrical without any distinctive head and neck. It projects medially at a right angle from the shaft and fits into a perforated acetabulum of up to 1.5 times its diameter. As a result, the hip joint is stable and fully congruent during parasagittal motion, permitting a wide range of flexion and extension, but very little abduction and adduction. The joint becomes quickly incongruent as the femur is abducted or adducted from the parasagittal plane of travel (Hotton, 1980).

The hip joint of pterosaurs is more mobile and profoundly different from that of theropods, but is reminiscent of that of mammals, especially of humans, allowing a wide range of adduction and abduction in the vertical plane. The great degree of vertical excursion of the femur is linked to its dual role in terrestrial and aerial locomotion. The controversy about terrestrial locomotion in pterosaurs has focused primarily on the stance and posture of the hindlimb. Using the ventral symphysis of *Anhanguera* as a guide, the acetabulum faces laterally and somewhat ventrally. The femoral component is a well-defined, spherical head, which is distinctly separated from the shaft by a narrow, non-articular neck at an obtuse angle of 160° (Fig. 8). The head forms a ball and socket joint with the close-fitting, shallow and imperforate acetabulum. The hip joint is fully congruent and stable throughout an enormous range of abduction and adduction, flexion and extension, and rotation, and perhaps was only constrained by soft tissue. When the femur is articulated with the acetabulum, the most stable position of the femur is in the horizontal direction in the wing plane, making an angle of 90° in relation to the fore and aft axis (Fig. 8A). From this position the femur can be moved forward and backward in the horizontal plane, sweeping an angle of 45°–125° relative to the fore and aft axis. This is the position of the femur that was adopted in flight, which permitted adjustment of the flight membrane.

In articulation with the sacrum, the pelvis of *Anhanguera* was closed ventrally with a pubo-ischiadic symphysis. In this orientation, there is no overhanging acetabular rim to support the femur in a vertical position. The femoral head is completely disarticulated from the acetabulum in the vertical position. We estimate that the amount of adduction possible at the hip joint ranges from 90° (horizontal) to 155° (near vertical) relative to the vertical axis of the body (Figs. 6, 8C). This observation is consistent with the range advocated by Wellnhofer (1985) and Unwin (1987), thus giving the femur a near-erect stance. The hindlimbs were used in a nearly, but not fully, vertical position, when walking on land.

The femur could be protracted and retracted considerably from this near-erect position but such a wide range of motion was not used during terrestrial locomotion (Fig. 8B). The unusual elongated forelimb defines the gait and restricts the forward and backward swing of the femur during terrestrial locomotion. The femur would assume a more vertical position to gain height with a limited forward and backward rotation to synchronize with the forelimb motion. In this aspect, the quadrupedal locomotion of pterosaurs is similar to that of primates, as discussed later. The cranial elongation of the iliac process in pterosaurs provided a

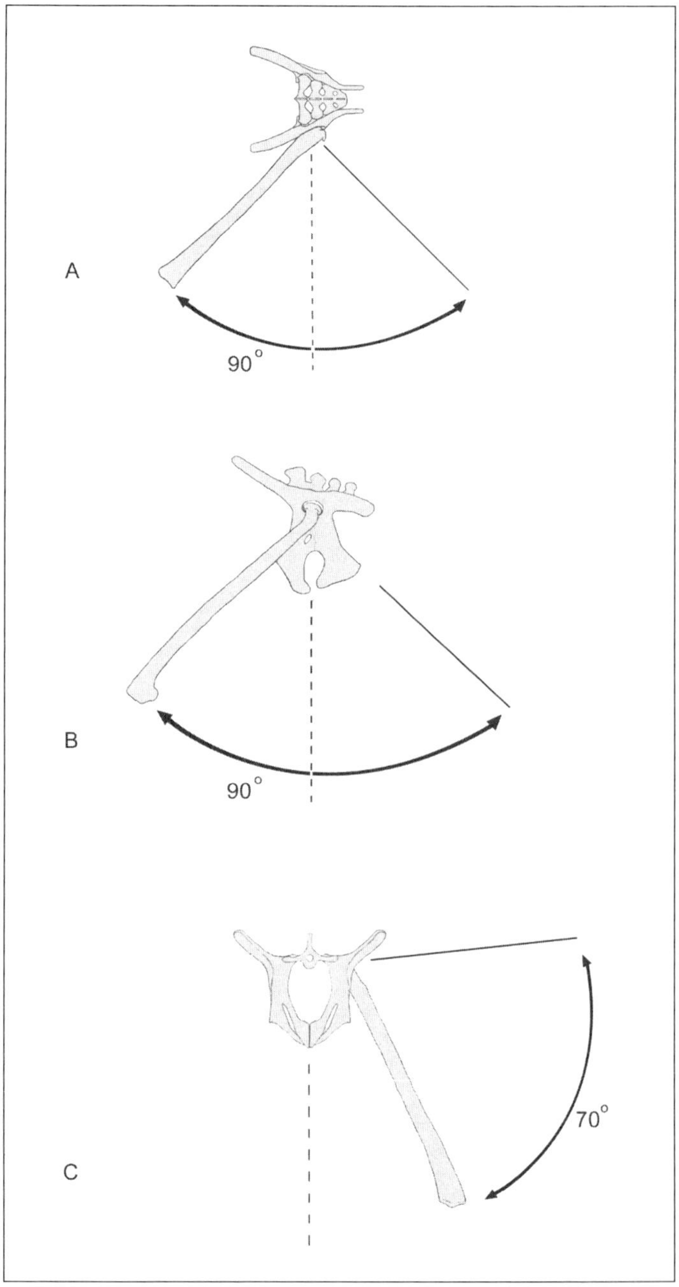

Figure 8. Reconstructed range of motion of the hip joint based on manipulations of three-dimensional cast skeleton *Anhanguera piscator* (TTU P10363). A: Dorsal view. B: Left lateral view. C: Cranial view.

large area for the attachment for M. iliotibialis cranialis, the main femoral protractor. However, the retraction of the femur was somewhat limited in pterodactyloids because of the loss of the tail. The smaller number of caudal vertebrae, reduced transverse processes, distal specializations of the tail, and the reduced fourth trochanter indicate the reduction or loss of the caudofemoralis muscle similar to the condition of modern birds (Gatesy, 1990). Unlike birds, pterosaurs had no antitrochanter locking device to keep the femur in a near-horizontal position. Movements at the hip, rather than at the knee, account for most of the foot displacement in pterosaurs. It is likely that the femur had a limited arc of protraction and retraction during walking. In the long-tailed "rhamphorhynchoids," the M. caudofemoralis was large and important, allowing a wide range of femoral retraction.

Knee Joint

The knee was a simple hinge joint where the bicondylar end of the femur rotates around the saddle joint of the crus formed by the tibia and splint-like fibula. The hinge motion is primarily around the horizontal axis, permitting a wide range of flexion and extension of the crus. This is the primary joint during terrestrial locomotion for birds. In pterosaurs, the tibia is much longer and more slender than the femur, and the fibula is considerably reduced as in birds. The longer tibia would confer a longer stride. This unusual limb proportion between the femur and the tibia led Padian (1983) to believe that pterosaurs were bipedal cursors. Recent evidence supports the notion that pterosaurs were quadrupedal most of the time, but assumed a bipedal posture during takeoff and landing.

Ankle Joint

The metamorphosis of the ankle joint of pterosaurs during development is interesting. In juvenile specimens, the proximal row consists of astragalus and calcaneum which are both reduced in length proximo-distally to become essentially only caps over the lower ends of the tibia, with the loss of the calcaneal tuber. Kellner and Tomida (1993) described the articulation between the astragalus and calcaneum as a "crocodile-reverse" joint (sensu Chatterjee, 1982) where the calcaneum has a convex surface that fits into the concavity of the astragalus. This reverse arrangement of ankle joint supports its inclusion in the ornithosuchian lineage. In adult specimens, these two bones are fused with the tibia to form a rolling hinge joint at their distal condyles as seen in modern birds. The ankle joint becomes a simple mesotarsal hinge between the proximal and distal rows of tarsal bones, permitting flexion and extension of the pes.

Pedal Stance

Padian (1983) argued that the metatarsals of pterosaurs were compact and closely appressed to each other so that only the toes contacted the ground during terrestrial locomotion in digitigrade fashion similar to the manner of theropods. However, in theropods, there are three functional digits where the central metatarsal is the longest and lateral members are shorter. In contrast, the inner four metatarsals of pterosaurs are subequal in length and do not show any tridactyl trend. This is not a normal digitigrade type of foot. Most workers (Unwin, 1988a; Wellnhofer, 1991a; Bennett, 1997a) preferred a plantigrade pes for pterosaurs. Discovery of a beautiful, three-dimensional and articulated pes of *Dimorphodon* has resolved the pedal stance of pterosaurs and supports a plantigrade interpretation (Clark et al., 1998). The plantigrade stance is consistent with the pterosaur trackways that show impressions of the entire sole of the foot (Fig. 9) (Lockley et al., 1995).

Trackways

The fossil trackways offer a positive evidence for the posture and gait of an extinct animal. They tell us immediately whether an animal is large or small and whether it is bipedal or quadrupedal. They tell us about the shape and position of the manus and pes and the number of toes. They also indicate whether the trackmaker walked erect or with a sprawling gait. Thus pterosaur tracks, if properly identified, would settle the controversy about the posture and locomotion. Unfortunately, trackways assigned to pterosaurs remain highly controversial but are gaining wide acceptance in recent times.

Stokes (1957) erected pterodactyl trackway to the ichnogenus *Pteraichnus* from the Jurassic Summerville Formation of northeastern Arizona. The trackway consists of nine consecutive steps showing quadrupedal progression. The manus print reflects an unusual digitigrade adaptation where the tridactyl manus and asymmetrical digits (I-III) are directed laterally, not cranially. The pes print has a distinctive V-shape with a sharply pointed heel and flat sole, indicating a plantigrade stance. The four diverging toes (I-IV) are subequal in length and oriented forward. The manus and pes prints are subequal, about 75 mm long, and are equidistant from the midline of the trackway (Fig. 9A).

Padian and Olsen (1984) disputed this interpretation of Stokes and suggested that *Pteraichnus* had been made by a small crocodilian, a conclusion implying that no pterosaur tracks are known (Unwin, 1989; Padian and Rayner, 1993). However, Lockley et al. (1995) rejected the crocodilian interpretation and resurrected the pterosaurian affinity of *Pteraichnus*, proposed by Stokes. They also described several trackways of *Pteraichnus* from different parts of the globe. They observed that the manus tracks are usually deeper than the pes tracks implying that the pterodactyl carried a majority of the body weight on the front limbs. In some trackways, only the manual prints were preserved (Fig. 9B). Lockley et al. (1995) postulated that the trackmaker was buoyant and was feeding in shallow water, while supporting the body weight mainly on the forelimbs. They concurred with Stokes that the manus impressions represent II-III-IV digits rather than I-II-III, and that the animal was quadrupedal, semierect, and plantigrade.

Mazin et al. (1995) described possible *Pteraichnus* tracks from the Upper Jurassic of France in which the tridactyl manus

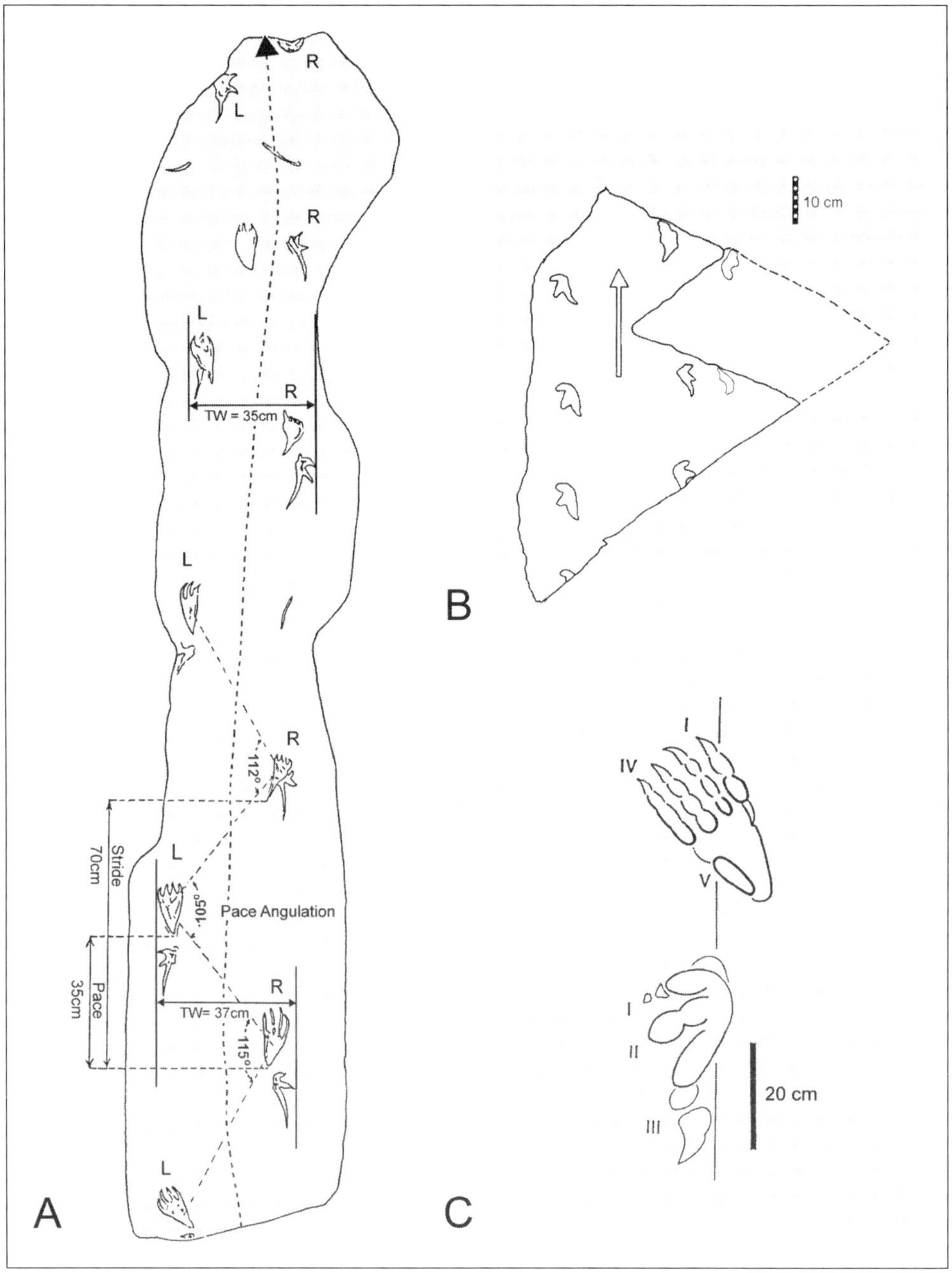

Figure 9. A: Pterosaur trackways *Pteraichnus saltwashensis* from the Jurassic Morrison Formation of Arizona. Note manual and pedal track width (TW) are narrow and similar dimensions, suggesting that the trackmaker was walking quadrupedally with a digitigrade manus and plantigrade pes; both forelimbs and hindlimbs were adducted closer to the body; manus track are typically deeper than the pes tracks, suggesting that the majority of weight was supported by the front limbs; arrow indicates direction of progression (modified from Stokes, 1957). B: *Pteraichnus* trackway from the Jurassic Sommerville Formation of Utah showing only manus impressions; the lack of pes impressions may indicate the unusual feeding strategy of pterosaurs in shallow water (see Fig. 36) where the trackmaker was buoyant and walking bipedally on its forelimbs on the floor of the basin in search of food (after Lockley et al., 1995). C: Details of pes and manus tracks of *Pteraichnus stokesi* from the Jurassic Sundance Formation of Wyoming showing the numbering of digits; note manus digits were directed laterally whereas pes digits more forward during walking (after Bennett, 1997a).

often shows an additional impression of metacarpal IV, pointing caudomedially (Fig. 9C). If this interpretation is correct, the three small digits actually represent I-II-III. The manus print requires a lateral torsion of the hand, a consequence of long-axis rotation of the humerus laterally during wing folding.

Bennett (1997a) interpreted that the manus print of *Pteraichnus* represents digits I-III. Using the trackways as a guide, he reconstructed the step cycles of *Pterodactylus* and concluded that the posture of the animal was erect rather than sprawling or semierect. We concur with Bennett about the digital numbering of the manus print (I-III).

Quadrupedal Walking

Birds have acquired two independent and specialized modes of locomotion, flying with the forelimbs and walking with the hindlimbs. The wing and tail are active during flight; the hindlimbs take no part in flight except during takeoff and landing. Similarly, when the hindlimbs are employed in walking, running, or perching, the forelimbs are folded up tightly against the body. In pterosaurs, there was no division of labor between the forelimbs and hindlimbs during aerial and terrestrial locomotion. Instead, both limbs were used during flying and walking. As a result, they were mobile and versatile. Unlike bats, the limbs in pterosaurs were held in near-parasagittal position. In comparison to the forelimb, the hindlimb appears to be a small structure in pterosaurs. This may be an illusion because the forelimbs are hypetrophied for flight; the legs are well sized with respect to the torso. The limb ratio often reflects the posture of the animal. In pterosaurs, the forelimb gait is fully digitigrade as it incorporates an extra segment, the metacarpals, whereas the hindlimb is plantigrade. Because of this postural disparity, the forelimb is proportionately much longer than the hindlimb (Table 2). There is a progressive increase of the length of metacarpal IV in relation to the radius in pterosaurs. In "rhamphorhynchoids," the metacarpal is shorter than the radius; in Jurassic pterodactyloids, it is longer than half of the radius; and in Cretaceous forms such as *Pteranodon* and *Quetzalcoatlus*, it is longer than the radius. Elongation of McIV probably signals a dramatic change in terrestrial locomotion from "rhamphorhynchoids" to pterodactyloids, a posture associated with a more upright vertebral column. The forelimb/hindlimb index [(humerus + radius + McIV) × 100/(femur + tibia)] is variable across the phylogeny, ranging from as low as 113 in *Quetzalcoatlus* up to 215 in *Pteranodon*. The quadrupedal locomotion of pterosaurs is somewhat analogous to the knuckle walking of apes with a steep spine angle; however, unlike the apes, the fingers were not folded but were extended and directed sidewise for additional support on the ground (Fig. 7). It is likely that the limbs of pterosaurs moved in a near-parasagittal plane. In pterosaurs, the center of mass lay between the two girdles when on the ground. As the animal changed the configuration of its limb elements during locomotion, the position of its center of mass changed from one instant to the next. The "rhamphorhynchoids," with their long bony tails, could possibly balance the body in a bipedal stance by shifting the center of mass backward towards the hip. For pterodactyloids, however, bipedal locomotion was far more difficult as the spine was held at a high angle from the horizontal position. They not only lost the bony tail, but they also lacked the expanded synsacrum seen in birds. Moreover, they had an enormous but light head to support in front of the body; the backwardly directed wing might have acted as a counterbalancing device like the two horizontal poles used by the acrobats during walking on high wire. The center of mass was located close to the shoulder girdle. Lockley et al. (1995) noted that manus prints are deeper than the pes tracks, indicating that the pterosaurs had more weight to support on their forelimbs. It is likely that in pterodactyloids, the body weight was supported by fore and hindlimbs during normal quadrupedal progression (Fig. 7).

Although we concur with the locomotory animation of pterosaurs by Henderson and Unwin (1999) in a broad sense, there are some anatomical inconsistencies. For example, Henderson and Unwin did not account for the automatic and synchronized flexion and extension of the elbow and wrist joints in pterosaurs. As the elbow joint is flexed during locomotion, the wrist joint is also flexed in pterosaurs. In Henderson and Unwin's animation, the wrist joint always opens at 180° even when the elbow joint is flexed. Henderson and Unwin did not consider the automatic rotational and ventral component of the wing digit during flexion that would bring the wingtips closer to each other. The known footprints do not support the forward orientation of the manus in Henderson and Unwin reconstruction, because the manus footprint is directed sidewise in pterosaurs (Lockley et al., 1995). Moreover Henderson and Unwin did not show bipedal locomotion of pterosaurs at all.

Bipedal Running

Padian (1983) argued forcibly the bipedal stance of pterosaurs. Like many modern quadrupedal lizards and primates, it is likely pterosaurs could rear up on their hindlimbs and momentarily become bipeds during rapid locomotion (Fig. 10). Chimpanzees occasionally walk bipedally with their knees bent and back sloping, especially when carrying things with their hands.

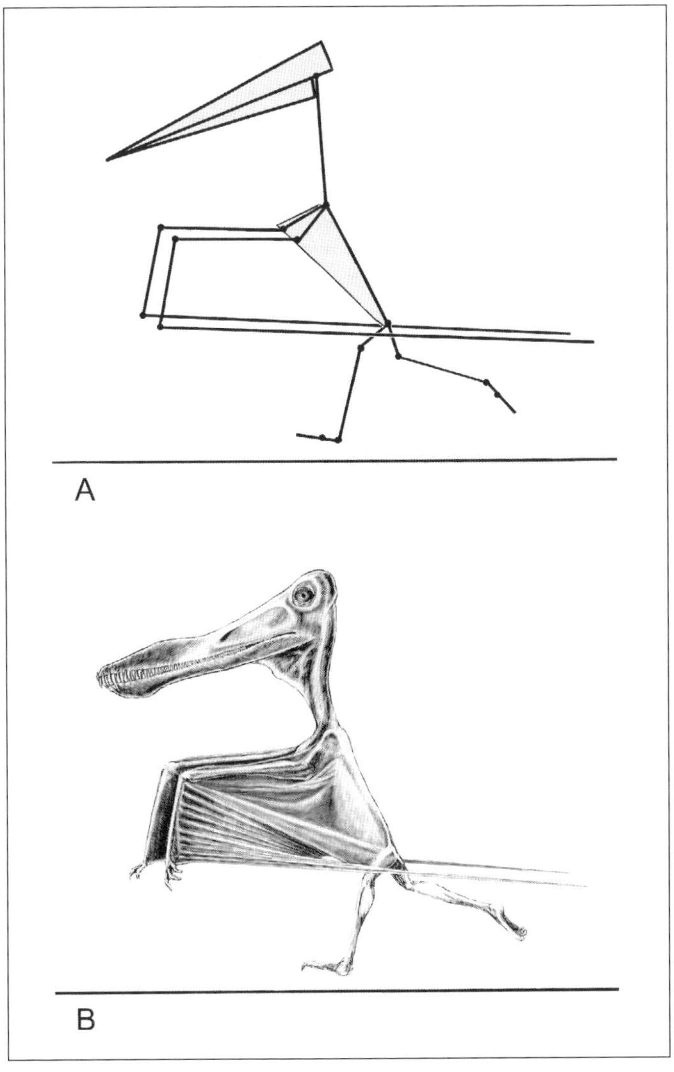

Figure 10. A: Stick diagram showing the bipedal running pose of *Anhanguera*. B: The same with flesh reconstruction.

Unlike the forelimbs, the hindlimbs of pterosaurs could be rotated forward and adducted to a position close to the body to allow a bipedal stance. Since the spine is at a steep angle in the quadrupedal pose, the bipedal posture can be easily assumed by keeping the sacrum at 60° above the horizontal, as Bennett (1997a) proposed. In this position, the wing digits can be projected considerably backward to compensate the body weight so that the center of mass can be positioned above the pes as in primates. In bipedal running, bent knees would confer speed. Although the bipedal position is unstable, the caudally directed folded wings could balance the body. During the transition from walking to running, there would be a drastic change of gait from quadrupedal to bipedal mode. During walking, each pes is on the ground for more than half the stride, so there are stages when both pedes are on the ground simultaneously. In contrast, when running, each pes is on the ground for less than half a stride, so there are stages when both pedes are off the ground. At these stages the body is highest in midair when it's kinetic and potential energies are also highest. At each footfall, these energies are lost and returned in an elastic recoil (Alexander, 1992). Moreover, during the transition from walking to running, the stance changes from plantigrade to digitigrade mode, making less contact with the surface to provide rapid footfall and increased stride. Other anatomical features such as the high tibia/femur ratio, reduction of the fibula, and the mesotarsal ankle are also associated with facultative bipedality (Padian, 1983). Bipedal running would be essential for a short burst during "takeoff" and "landing" with part of the weight being supported by wings flapping (Fig. 10).

WING DESIGN

Comparative Morphology of the Wing

The primary function of the wing is flight. The wing morphology of flying tetrapods partly reflects their flight style, heritage, and environments in which they live. In pterosaurs, birds, and bats, the wing is supported by elongated and modified forelimbs. Specialized flight surfaces such as membranes or feathers give the wings lightweight and mechanical integrity. The wings of birds, pterosaurs, and bats evolved convergently for flight adaptation from three separate lineages. Both pterosaurs and birds belong to the clade Archosauria, whereas bats are grouped within the Mammalia. As a result, pterosaurs and birds show closer phylogenetic similarity, as reflected in their anatomy and lifestyle. On the other hand, pterosaurs and bats evolved membranes as a flight surface. Consequently, they share functional similarity in the wing design through convergent evolution. Because of these phylogenetic divergences but functional similarities, pterosaurs were neither exactly like bats nor exactly like birds. They had evolved their biology and adopted the terrestrial and aerial mode of life for 160 million years in a uniquely pterosaurian way, quite different from those of birds and bats.

The spanwise lengthening of the wings of birds, bats, and pterosaurs, especially in the distal section, is the most obvious adaptation of active flight and it results in reduced induced power requirements (Fig. 11). Pterosaurs and bats achieved this by lengthening the fingers internally to support the wing membrane. Birds acquired the long span externally by attaching long primary feathers that are sufficiently rigid. As a result, the bones of the forelimb in birds do not stretch to the tip of the wing but are restricted proximally to support the base of the feathers. The primary feathers, attached to the manus, are responsible for providing forward thrust. The secondary feathers, attached to the ulna, provide lift. There are also a few tertiary feathers on the humerus, as well as three or four feathers on the thumb, called the alula, which are used to prevent stalling at low speed. The flight feathers overlap in the wing in such a fashion that they can be folded or extended easily in conjunction with the flexion and extension of the joints in the wing elements. When the wing is spread, the flight feathers are automatically set in place, at an angle to the wing bones to which they are attached (Chatterjee, 1997). The central rachis of a flight feather is a strong but flexible hollow tube to constantly accommodate twisting and bending, in order to maintain the proper angle to the airflow without snapping off. The hand bones in birds are greatly reduced and fused, opposite to the condition in bats and pterosaurs. The trailing edge of avian wings is attached to the side of the body, so that the hindlimbs remain completely free during flight and terrestrial locomotion. The wings are folded automatically at the elbow and wrist joints in birds to reduce wingspan during the upstroke. Since the forelimbs can be pressed against the body, a bird can reduce its wingspan by a larger amount than bats or pterosaurs can (Pennycuick, 1986, 1988).

Both bats and pterosaurs have wings of skin membrane, which is partitioned into four panels: the forewing or propatagium, the lateral finger membrane or dactylopatagium (= actinopatagium), the medial arm membrane or plagiopatagium (= tenopatagium), and the tail membrane or uropatagium (sensu Yalden and Morris, 1975). In pterosaur literature, the main wing is simply called the brachioptagium. In our discussion, the brachiopatagium is further subdivided into two components, the plagiopatagium and dactylopatagium, because of their different morphology, function, and distribution of structural fibres (actnofibrils). However, bats and pterosaurs support their skin membrane in different fashion. In bats, the membrane itself is very thin, perhaps only 0.03 mm thick in small bats. It is elastic and consists of two layers of skin, each with a thin epidermis and a thicker dermis. It incorporates a number of fine blood vessels and small bundles of muscles and is reinforced by elastic fibers, which run perpendicular to the fingers (Vaughan, 1966). The outer four fingers (II–V) form an inner support of their dactylopatagium; the thumb is free, reduced, and tipped with a claw. A ligament runs from the second to third digit to form the leading edge. The wing membrane is spread between the fingers and the feet, so that it remains taut during flight. A cartilaginous spur, the calcar may project proximally from the ankle to support the uropatagium. The propatagium is supported by a special muscle, the occipito-pollicalis, running from the skull to the thumb (Yalden and Morris, 1975).

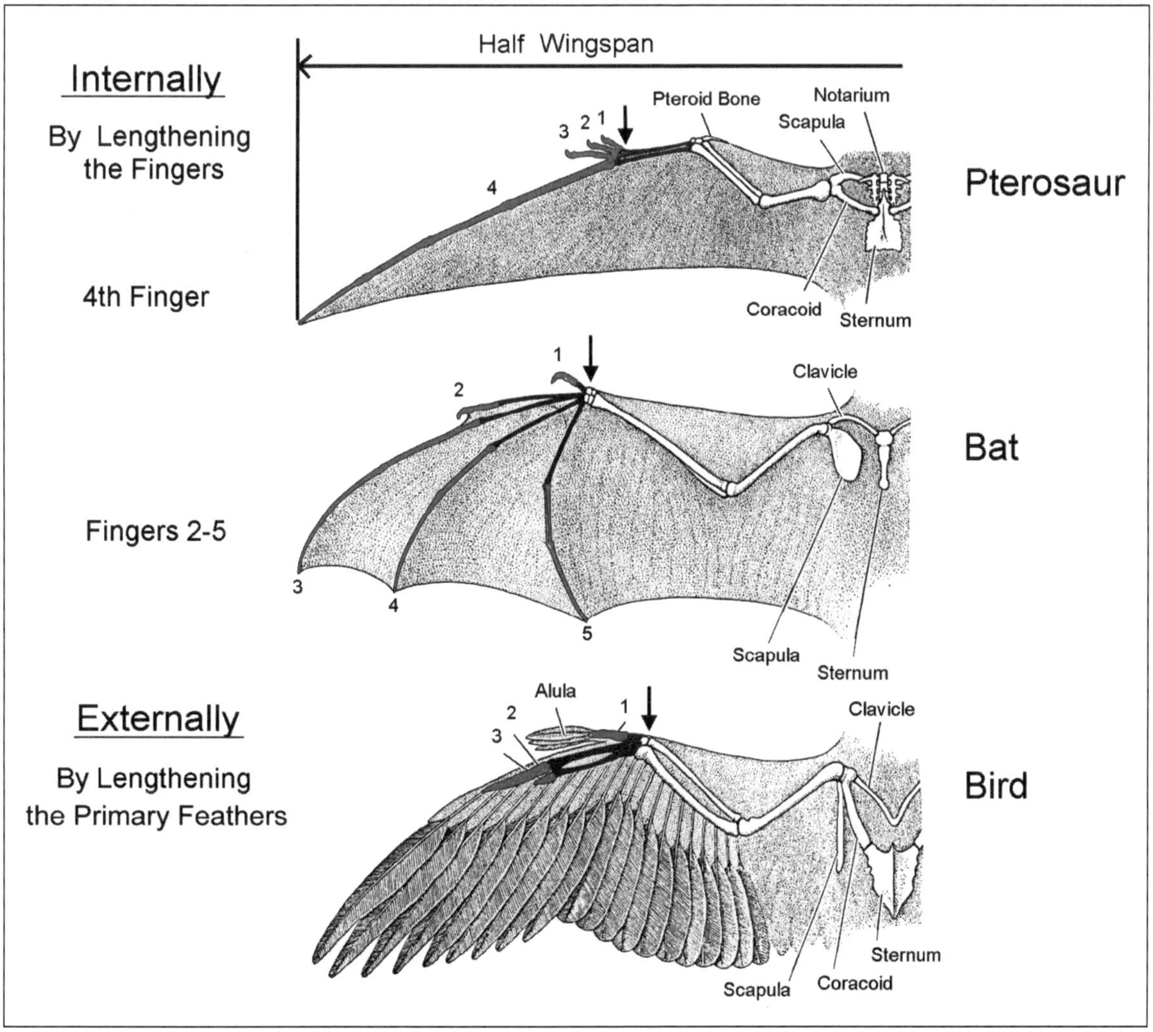

Figure 11. Diagrammatic comparisons of the wings of three groups of vertebrate flyers. The spanwise lengthening of the wings, especially the distal section, is the most obvious adaptations of active flight in these groups. Pterosaurs and bats have achieved this by lengthening the fingers (shown by gray) internally to support the wing membrane. Birds have acquired the long span externally by attaching long primaries feathers. In birds and bats, the main wing folding joint is at the wrist, whereas in pterosaurs, it lies at the knuckle (modified from Langston, 1981).

The amount of camber is a major factor in determining the ability of a wing to develop lift. Bats have the capability to vary the camber by lowering the thumbs and legs. Because the long fingers can spread and fold the wing like the ribbing of an umbrella, this allows a high degree of control of camber and angle of attack of the wing. As a result, bats have broader wings with low aspect ratios but greater maneuverability and the ability to fly slowly. By moving their feet and fingers, bats control the shape and tautness of the wings (Vaughan, 1966; Yalden and Morris, 1975).

The skin membrane of pterosaurs superficially resembles that of bats with four distinct panels (Fig. 12C). The forewing or propatagium is supported by the pteroid bone, unique to pterosaurs (Fig. 11). It is a long splint, projecting from the wrist and pointed toward the shoulder to support the propatagium. The homology of the pteroid bone is controversial; most probably it represents a modified first distal carpal (Unwin et al., 1996). Birds and bats both have a propatagium, but in each case an elastic membrane supports it, not by a bone. The dactylopatagium in pterosaurs is supported by one hyperelongated fourth finger. This flight digit is much thicker and longer than the fingers of the bat. The extreme elongation of the fourth finger makes the pterosaur wing narrow with a high aspect ratio and a low wing loading. The first three fingers in pterosaurs are normal, with a typical phalangeal formula of 2-3-4, and bear sharp claws; they do not take part in flight. However, they might have played important roles for climbing vertical surfaces such as cliffs and tree trunks; they were also used to support weight during terrestrial locomotion. The fourth finger is the

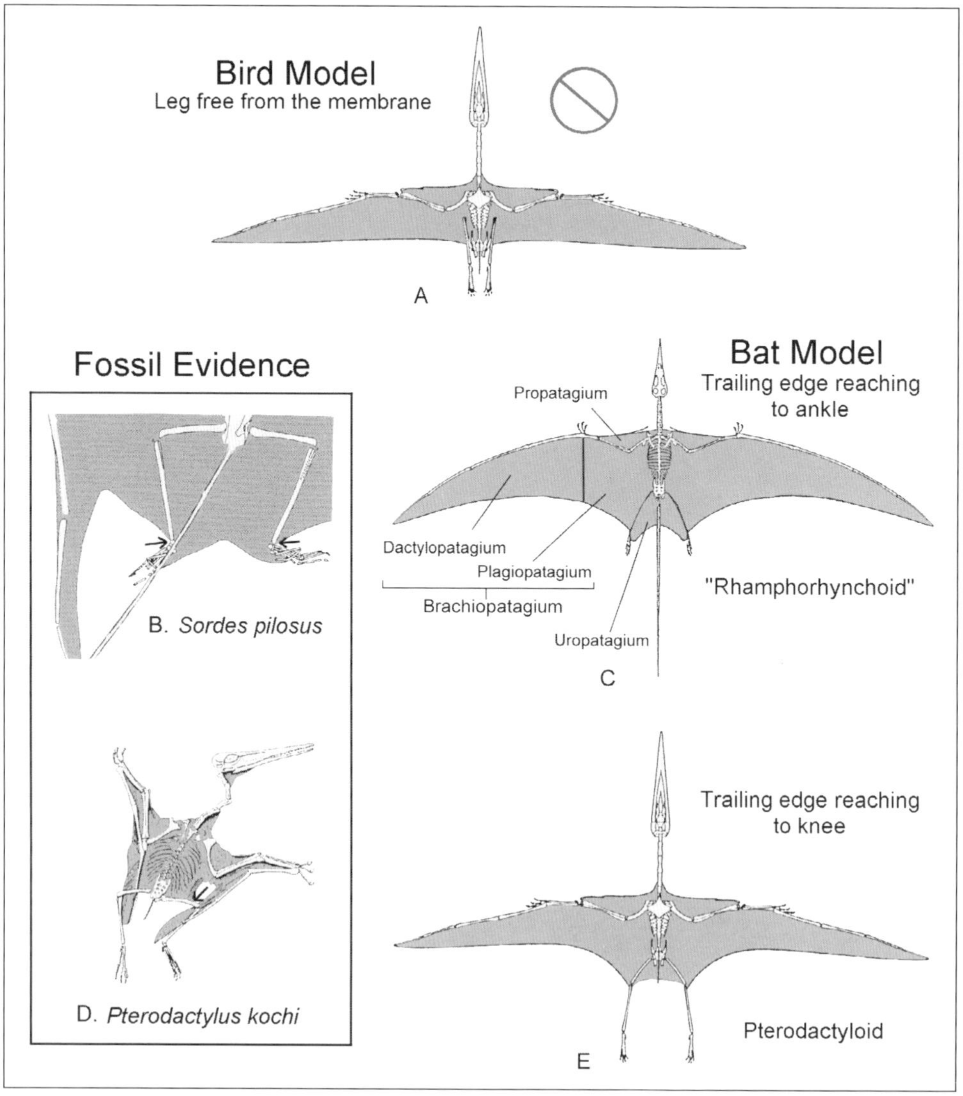

Figure 12. Wing planforms for two groups of pterosaurs: A. Narrow planform reconstructed for the leg being free of the membrane (after Padian, 1983); however, fossil evidence contradicts this view. B: *Sordes pilosus*, a Jurassic "rhamphorhynchoid" from Kazakhstan in which exceptionally preserved wing membranes show attachment with the hindlimb near the ankle (shown by arrow) (modified from Unwin and Bakhurina, 1994). C: Broad wing planform of "rhamphorhynchoid" based on *Sordes*, showing four panels of wing membrane—propatagium, dactylopatagium, plagiopatagium, and uropatagium. D: *Pterodactylus kochi*, a Jurassic pterodactyl from Germany (Vienna specimen) indicates that the wing membrane extended to the upper leg near the knee joint (shown by arrow) rather than down to the ankle (modified from Wellnhofer, 1991a). E: Narrow wing planform of pterodactyloids based on the Vienna *Pterodactylus*.

longest in the series consisting of four phalanges, but it lacks the ungual. The fifth finger is lost in pterosaurs.

The spanning structures of the wings of birds, bats, and pterosaurs can be divided into three segments at three flexible joints: the proximal, medial, and distal structures (Fig. 11). The proximal segment (humerus) encompasses the portion between the shoulder and the elbow; the medial segment (radius/ulna) lies between the elbow and the wrist; and the distal segment (metacarpals/phalanges) contains the portion between the wrist and the tip of the wing. In birds and bats, the main wing-folding joint is at the wrist, whereas in pterosaurs, it is situated at the knuckle (Fig. 11) between the fourth metacarpal and the wing phalanx (Hankin and Watson, 1914). Thus, an additional segment, composed of fourth metacarpals, has been added in the medial spanning structure in pterosaurs; as mentioned before, this segment is important during terrestrial locomotion. The distal segment consists of the portion between the knuckle and the tip of the fourth digit. The lateral shift of the distal segment of the spanning structure, from the wrist to the knuckle joint, is a unique feature in pterosaurs. The knuckle joint is the main wing-folding mechanism in pterosaurs. Pterosaurs are distinguished from bats and birds by the relative shortness of humerus and forearm and the hyperelongation of the fourth digit.

Wing Geometry

The pteroid bone and its tendon, the conjoined four metacarpals, and the fourth finger form the leading edge of the pterosaur wing—the wing spar (Figs. 11, 12). A number of workers have speculated that there may have been a tendon, connecting the tip of the pteroid bone to the coracoid for tension, which would form the leading edge of the propatagium. However, there is no hard evidence of such a tendon in pterosaur. Frey and Riess (1981) and Pennycuick (1988) argued that the pteroid bone was highly mobile in pterosaurs and could change the geometry of the forewing during flight, as in bats. Wellnhofer (1991a) argues that the pteroid bone appears to be too delicate to withstand such an enormous stress while changing positions. Since the pteroid bone is so mobile, it probably controlled the forewing like the flap of an aircraft as a brake, especially during takeoff and landing. It may be important in generating lift, too. The propatagium of pterosaurs is relatively smaller than the equivalent part of the bat membrane. In bats, the large membrane along the front of the wing can be used as a flap, another adaptation to low-speed flight (Lighthill, 1975).

Distal to the propatagium, the wing spar of pterosaurs, formed mainly by the fourth metacarpal and wing phalanges, is a hollow tube, trussed inside to make it light and strong to withstand great strains during flight. The flight digits are hyperelongated, rounded to oval in cross-section, gradually compressed and tapered distally. They are bent backwards like a bow to give a nice arch of the wing spar.

The nature and extent of the trailing edge of the plagiopatagium are currently the subject of much disagreement. Part of the controversy stems from the poor preservation of the wing membrane. Very few specimens actually show the contour and extent of the caudal margin of the membrane clearly and unambiguously. Often, the margin is diffuse and merges into the matrix without any visible boundary. Since pterosaurs are extinct, most attempts to reconstruct their wing planforms are based on living analogs such as bats and birds. The extent of the trailing edge of the membrane has been correlated with the stance of pterosaurs. For example, if the membrane were attached to the hindlimb, as is traditionally portrayed (Soemmerring, 1820; Marsh, 1882; Bramwell and Whitfield, 1974; Wellnhofer, 1991a), the terrestrial posture would be sprawling, clumsy, and quadrupedal as in bats. On the other hand, if the membrane were narrow and attached to the body, keeping the hindlimbs free, the animal would be less cumbersome during terrestrial locomotion. This implies a more birdlike stance, with an erect and bipedal posture and a digitigrade pes (Padian, 1979, 1983; Brower, 1983; Rayner, 1988; Hazlehurst and Rayner, 1992a; Padian and Rayner, 1993). Here we examine the extent of the trailing edge. Which is the more likely analogy, the bird model or the bat model?

In the bird model (Fig. 12A), the membrane is trimmed to a narrow outline, where the trailing edge is attached to the sides of the body keeping the hindlimbs free. In flight, the hindlimbs would be folded up against the body as in birds (Padian, 1983; Brower, 1983; Rayner, 1988; Hazlehurst and Rayner, 1992a; Padian and Rayner, 1993). The bird model for the pterosaur wing has been heavily criticized in recent years by Pennycuick (1988), Wellnhofer (1991a), Bennett (1997b), and Unwin and Bakhurina (1994) on several grounds. First, the model is based on negative evidence. No specimen shows a clearly defined trailing edge, attached unambiguously to the sidewall of the body. Second, the long trailing edge would remain free and dangling during flight without any mechanism of tensing the membrane. From a biomechanical point of view, this is undesirable, especially for large pterodactyloids. Without leg attachment, there would be no angle of incidence control and alteration of the wing planform. As a result, wing adjustment during flight would be difficult, especially for large pterodactyloids. Third, the narrow, "birdlike" wing without any leg attachment would make the control of pitch and camber more difficult. Fourth, it would decrease the wing area slightly, thus increasing the wing loading and bending stress on the spar. Pterosaur flight performances may not depend much on whether the wing is broad or narrow, because differences in wing loading canceled out differences in aspect ratio (Brower, 1983; Hazlehurst and Rayner, 1992a). However, the involvement of hindlimb in manipulating the wing planform appears to be an important aspect in their flight styles. Some proponents of the bird model have subsequently accepted the modified version of the bat model to include the patagium attaching to the femora, but with the femora still folding up forward in flight as in birds (Hazlehurst and Rayner, 1992a).

Bennett (1987) modified the bird model and suggested an alternative point of attachment for the trailing edge. Instead of trunk attachment, Bennett speculated that the plagiopatagium in large pterodactyloids such as in *Pteranodon* was attached to the tail. He argued that the last segments of the caudal vertebrae consisted of pair of long and flexible rods possibly for the attachment of the plagiopatagium. By moving the tail and up and down, it can be used as a pitch control device. In this model, the forelimbs would be kept free as in birds. However, the short and slender tail of pterodactyloids appears to be too slender to support such a large wing membrane. Later, Bennett (1997a, 2000) recanted and accepted the bat model, where the tail could be used to support a small flap of uropatagium.

Here we prefer the bat model for pterosaur wings on the basis of fossil evidence with two planform variations that evolved and were refined through time: the broad-winged "rhamphorhynchoids" and the narrow-winged pterodactyloids (Fig. 12C, 12E). Modern bats show similar variations of wing configuration. In the "rhamphorhynchoid" planform, the plagiopatagium was extended to the ankle. In addition, there is a complete uropatagium stretching between the distal end of the metatarsal V and the tail, as revealed from a new specimen of *Rhamphorhynchus* (Tischlinger and Frey, 2002). An exquisitely preserved fossil of *Sordes* from the Upper Jurassic Karabastau Formation of Kazakhstan (Fig. 12B) clearly supports this "rhamphorhynchoid" model (Unwin and Bakhurina, 1994). The deposits in which it was found are lacustrine in nature. *Sordes* is a "rhamphorhynchoid" with a wingspan of 0.65 m. The specimen exhibits exceptional soft-tissue preservation, which authoritatively illustrates both the shape of the main wing membrane and individual fiber strands. The preservation displayed is better than that produced by the Solnhofen Limestone of Germany. Therefore, with regard to "rhamphorhynchoid" wing planform, *Sordes* currently provides us with the most satisfactory answer.

The flight membrane is frozen in the skeleton of *Sordes*. Within the angle of the humerus and forearm, a small triangular propatagium is present. Unwin and Bakhurina (1994) suggested that the pteroid bone manipulated this membrane. The preserved plagiopatagium originates along the rear edge of the forelimb and extends along the body from the shoulder to the hip and the cranial margin of the hindlimb as far as the ankle. During flight, the femur of *Sordes* pronated by 90°, and the legs are splayed sidewise in such a fashion that the knee joint flexed and extended in the plane of the wing. The pes is directed caudally. The fifth toe is very long, clawless, and composed of two phalanges, where the second phalanx articulated with the first to form a right-angle bend. This bone is often compared with an analogous calcar element of bat, which occurs medially for the attachment of a segment of the uropatagium that occurs between the legs. Unwin and Bakhurina (1994) argued that this structure does contain fibers and, therefore, is not displaced skin. It is a true uropatagium. The pes was pointed backward in *Sordes* from the normal position during flight so that the soles of the feet faced each other like a flying fox; as a result, the fifth toe occupied the position dorsally. In this position, it supported, tensed, and manipulated the uropatagium analogous with the calcar of bats. The high degree of mobility in the interphalangeal joint would facilitate these movements. *Sordes* also suggests that the pterosaur wing was structurally non-homogenous. The distal part of the wing (dactylopatagium) is composed of long, straight, and closely packed fibers, which suggest that this portion of the wing was rather stiff and strong. However, the proximal and medial parts of wing (propatagium, uropatagium, and plagiopatagium) were more pliable (Unwin and Bakhurina, 1994).

Several specimens of *Rhamphorhynchus* also show a caudally directed pes, where the fifth toe has been turned dorsally as in *Sordes*, possibly for the attachment of the uropatagium (Fig. 12B). The inclusion of the hindlimbs in the wing and tail membrane would add greatly to the control of wing camber and wing twisting (Norberg, 1985). The broad wings of "rhamphorhynchoids" were useful for hovering flight, as discussed later. Pennycuick (1986) pointed out that in some specimens of *Rhamphorhynchus*, both legs have been drawn up toward the wing, ankles first indicating that the ankles were attached to the wing membrane. If this interpretation is correct, "rhamphorhynchoids" walked on all fours. Pennycuick further speculated that an elastic ligament extending from the wingtip to ankle would keep the membrane taut. Recent discovery of a well-preserved *Rhamphorhynchus* specimen indicates the presence of a tendon that reinforced the trailing edge of the brachiopatagium (Tischlinger and Frey, 2002).

In the pterodactyloid planform, the wing became narrower (Fig. 12D, 12E). The trailing edge of the membrane is attached to the femur or knee joint, keeping the lower leg free (Wellnhofer 1988, 1991a, 1991b). This kind of wing design is beautifully preserved in the Vienna *Pterodactylus* (NHMW 1975/1756), which clearly indicates that the trailing edge of the membrane extended to the upper leg rather than down to the ankle (Fig.12D). However, the exact point of attachment of the membrane is not clear. In this specimen, the femur is extended backward, not under the body. This posture clearly contradicts the birdlike folding of the legs proposed by Padian (1983). Since the femoral shaft does not show any kind of ridge of protrusion for the attachment of the membrane, we speculate that the membrane was attached farther down at the knee joint, possibly at the lateral epicondyle of the femur, keeping the lower leg free. This arrangement would help the pterodactyloids to walk comfortably on land. In *Pterodactylus*, the fifth toe became small with a tiny phalanx, whereas Cretaceous pterodactyloids lost the fifth toe altogether. A small part of the uropatagium is probably attached between the tail and the medial epicondyle of the femur. In a Munich *Pterodactylus* (BSP 1937.1.18), the uropatagium is preserved medial to the right femur, indicating its presence. It would stretch from the knee joint to the tip of the tail. During flight, the hindlimbs were directed horizontally in the same plane with the wing membrane.

Padian and Rayner (1993) questioned the presence of a uropatagium in pterosaurs, because it would interfere with their bird-model of terrestrial locomotion. As discussed earlier, several specimens do show the impressions of the tail membrane. For example, in "rhamphorhynchoids" as in *Sordes* (Sharov, 1971; Unwin and Bakhurina, 1994) and *Rhamphorhynchus* (Wellnhofer, 1975; Tischlinger and Frey, 2002), the uropatagium is stretched between the fifth toe to the proximal segments of the tail. In this group, the fifth toe occupies the same topographic position as the calcar of bats and functioned similarly for supporting, tensing and manipulating the uropatagium (Unwin and Bakhurina, 1994). In pterodactyloids, the uropatagium appears to be narrower and attached to the femur near the knee joint and at the base of the tail. This planform of narrow uropatagium is seen in the Vienna *Pterodactylus* (Wellnhofer, 1970), *Pteranodon* (Bennett, 1987), and an unnamed Brazilian pterodactyl (Martill and Unwin, 1989). The apparent lack of fragile uropatagium in other pterosaurs fossils is probably the result of its taphonomic destruction.

Camber

Airplanes are usually designed with cambered wings, curved on the upper surface to produce lift. The wing of a pterosaur is constructed on similar aerodynamic principle. Not only is it streamlined to cut through the air with little drag, but it is also curved to produce lift. The nature of camber in pterosaurs is poorly understood because the mode of attachment of the patagium to the wing spar is not clearly preserved in any specimens. Wellnhofer (1991a) suggested that the wing phalanges of "rhamphorhynchoids" have a deep, longitudinal groove at the caudal edge of the first phalanx, probably for the insertion of the dactylopatagium. However, no such groove can be seen on the wing phalanges of pterodactyloids. Instead, there is a faint longitudinal ridge. One of the best-preserved pterosaur specimens with an intact wing is a *Rhamphorhynchus* specimen, which was first described by Zittel (1882). This specimen is commonly referred to as the Zittel wing, housed at the Bavarian State Collection Museum at Munich, Germany (BSP 1880.11.8). Padian and Rayner (1993) inferred from the Zittel wing that the dactylopatagium originated from the dorsal surface of the wing phalanges, rather than from the caudal groove. From the aerodynamic point of view, this interpretation is logical. The leading edge should merge smoothly with the upper and lower surfaces of the membrane without any break to avoid turbulence. In most reconstructions (Bramwell and Whitfield, 1974; Brower, 1983; Wellnhofer, 1991a), a thin, highly cambered airfoil was assumed for the pterosaur wing where the cylindrical leading edge ends sharply to the thin membrane of the wing. As the angle of attack and the lift coefficients are reduced, the flow separates from the leading edge beneath the wing with a consequent increase in drag, which becomes more severe as the angle of attack is reduced. The effect of the high drag seriously increases the sinking speed as flight speed is increased. It is not clear that this is a necessary condition for the wing aerodynamics of pterosaurs. Natural selection would make the wing more efficient in the highly successful pterosaur group. To overcome this turbulence, we speculate some sort of stretching that would partly flatten the wing camber at low angles of attack. It is likely that a thin sheet of fascia and muscle was attached behind the cylindrical flight digits and between the two layers of the skin membrane to streamline the wing (Fig. 13). The longitudinal groove of the wing phalanges of "rhamphorhynchoids" may indicate the insertion area for the muscle fibers. In pterodactyloids, these muscles might have been inserted on the caudal ridge of the wing phalanges. The effect of streamlining a circular cylinder by adding a tapered tail is well known in fluid mechanics (Kreider, 1985). Extension and tapering of the cylindrical object partially into the area of turbulence helps to maintain the smooth flow of air and prevents or reduces turbulence. A small circular cylinder has a drag coefficient of about 1.2 based on its frontal area, and that it is reduced by half if the cylinder is stretched in length to twice its diameter. Further stretching to four times the diameter reduces drag to about 30% of the unstreamlined shape. Streamlining or stretching is much more effective behind the cylinder than ahead of it. The effect of streamlining is to reduce turbulence and the width of the wake behind the cylinder. A combination of streamlining and flatter camber might be the best solution for pterosaurs. Recent wind tunnel flow visualization experiments have been made by DeLaurier (1989) on a

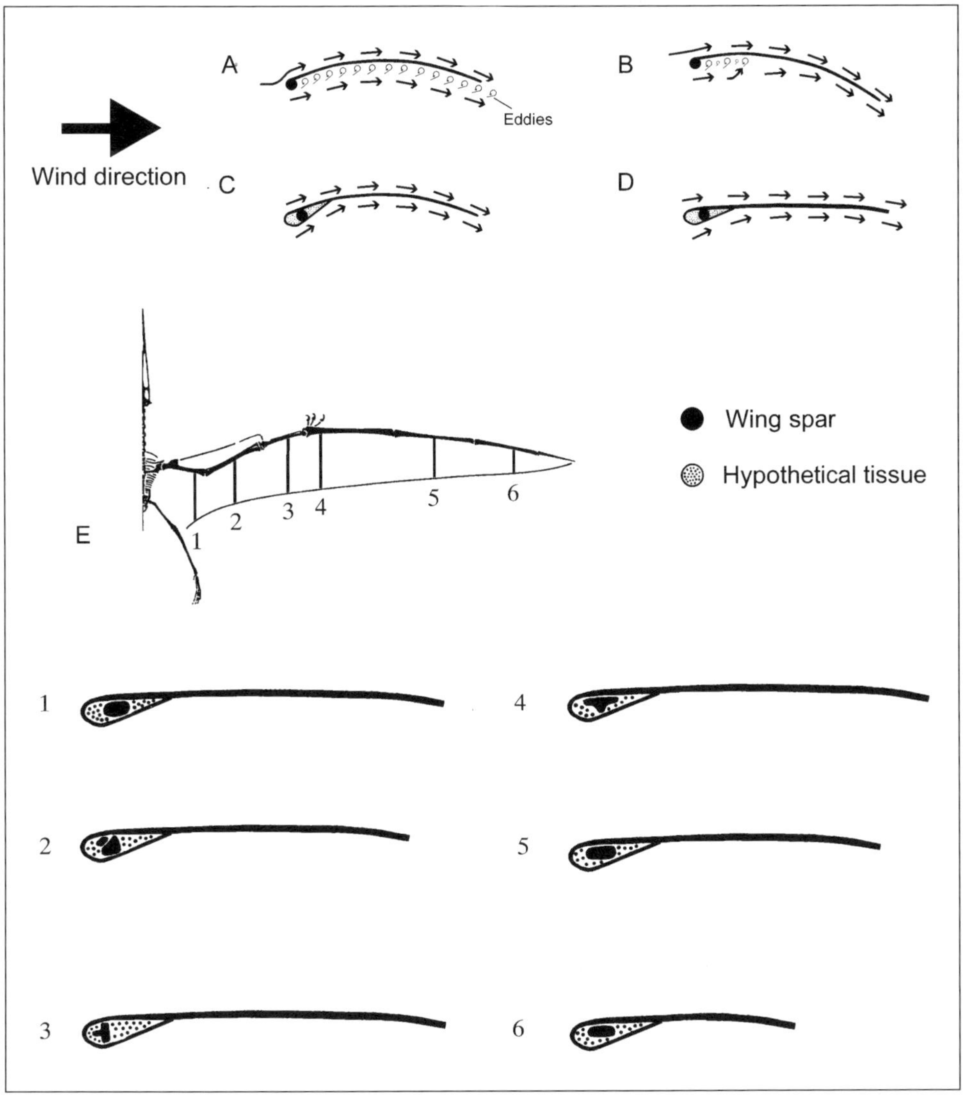

Figure 13. Camber and streamlining the pterosaur wing; wing profiles show the camber at idealized cylindrical section (A–D). A: Low-angle attack (high speed gliding); a thin, highly cambered airfoil was assumed for the pterosaur wing where the cylindrical spar would create turbulence behind it. B: The same profile at high angle of attack (slow gliding); drag decreases rapidly at higher angles of attack. C: The spar is streamlined by adding hypothetical tissue in front and behind the spar to reduce turbulent flow. D: The turbulent flow could be reduced further by flattening the camber; a combination of streamlined spar and flatter camber might be the best design for the pterosaur wing. E: Dorsal view of the wing of *Anhanguera* to show the locations of sections of the wing profiles as marked by 1–6; 1, section at the middle of the humerus; hypothetical tissues added both in front and behind the spar; 2, section at the middle of the radius-ulna; tissues added both in front and behind the spar; 3, section at the middle of the metacarpals; tissues added both in front and behind the spar; 4, section at the proximal region of first wing phalanx; tissues added both in front and behind the spar; 5, section at the second phalanx; tissue added behind the spar; 6, section at the third phalanx; tissue added behind the spar; 1–6 wing profiles enlarged six times from Figure E.

family of thin (curved plate) cambered airfoils at low Reynolds number (1.2×10^5) with and without leading edge fairings. Although Reynolds numbers for large pterosaurs in flight may approach 10^6 and may be expected to have some effect on flow separations and stalling, and although lift and drag forces were not measured in the experiment, some of the results are interesting and perhaps surprising. For instance, for sharp-edged cambered airfoils without fairing there is no serious separation of the flow on either surface at angles of attack from about –3° to 8° of small camber (5% of chord length), and from about zero to about 9° or 10° when camber was increased to 7.5%. Flow separations were actually worse with the addition of simple circular-nosed leading edge fairings, but from the experiments it would seem that an elliptic-nose fairing covering the bony leading edge spar, biased toward the upper surface could restore attached flow over the same angle range as that of the sharp-edged airfoil with no leading edge spar. Airfoils of higher camber were not tested, but the experiment suggests the advantage of evolving some form of leading edge streamlining. Airfoils with low camber are effective at producing lift at higher speeds and produce little drag (Fig. 13). It is likely that the camber of the patagium could be adjusted by elevation or depression of the pteroid bone and femur. The flight performance estimates discussed later are based on the assumption that the camber of the pterosaur wings is increased as the lift coefficient increases.

Wingtips

The long wingspans of pterosaurs and many seabirds are important in minimizing the lift-induced drag, which may be about half the total drag in cruising flight or in flat glides. Pterosaur reconstructions always show sharply pointed wingtips, because the wing membrane is attached to a single spar at the leading edge that also ends with a sharp tip. This sharply pointed wing design is undesirable from an aerodynamic point of view because it tends to stall at the tips, requiring "washout"—a reasonable tradeoff for high aspect ratio and low weight at wingtip to avoid this. However, the best-preserved Zittel wing of *Rhamphorhynchus* shows a blunt tip and this morphology has been observed in a great number of pterosaur specimens (Döderlein, 1929; Padian and Rayner, 1993). A rounded wingtip is better designed aerodynamically to avoid local flow-separation and stalling (Fig. 14). Local turbulence is reduced around a blunt wingtip. When compared to a sharp tip, a blunt tip is stronger if not excessively thinned. In bats, multiple wingtips at the trailing edge are pointed because the elastic membrane is stretched between the bony fingers. Bats do not have reinforcing fibers in the wing membrane, as did pterosaurs. In pterosaurs, the membrane is stiffened by actinofibrils that are bundled at the end to maintain a rounded tip geometry (Padian and Rayner, 1993). It is generally believed that certain seabirds with high aspect ratios have pointed wingtips, but do they really? On closer examination, the longest primaries in these birds tend to show rounded tips. For example, Burton (1990) provided some excellent plan view drawings of several high-aspect ratio seabirds from which we have selected two of the most pointed forms: the grey plover and wandering albatross (Fig. 14B, 14C). On these wing silhouettes, we have plotted black dashes to show the elliptic chord distribution that fits them closely, except for one or two tip feathers. We see a similar evolutionary pattern of the sail of a racing yacht at its distal tip (see America's Cup Museum, www.acmuseum.com). As shown in Figure 14D, it took one and half-centuries for racing yacht sail designers to modify the sharply pointed, low aspect ratio sail to more rounded version of higher aspect ratio for better aerodynamic performance.

Actinofibrils

Unlike the flexible bat wing, the wing membrane in pterosaurs is semirigid, with an arrangement of fine, parallel fibers radiating throughout the dactylopatagium. The arrangement of these fibers is analogous to the distribution of the feather shafts of the birds (Zittel, 1882; Padian, 1983). These fibers are oriented perpendicular to the direction of spanwise tension to maximize the strength and stiffness required during flight. The orientation of these fibers is at almost at a right angle to the orientation of the elastic fibers of bats. They continue medially to the plagiopatagium area, but they are less regular and fade away quickly. These fibers are beautifully preserved in the wing membrane of *Rhamphorhynchus* (Fig. 15A, 15B), first observed by Zittel (1882). The fibers are 0.05 mm thick with a uniform spacing of 0.2 mm (Wellnhofer, 1991a). Wellnhofer (1975) aptly coined the term actinofibrils, meaning "little ray fibers" to describe these structures. He reconstructed them as internal structures, which are roughly cylindrical and flexible rods, deeply embedded inside the skin membrane. However, Padian and Rayner (1993) argued that the actinofibrils were external, attached to the ventral surface of the patagium. The location of actinofibrils within the wing has been settled recently with discovery of an exquisitely preserved wing of *Rhamphorhynchus*. Tischlinger and Frey (2002) documented convincingly three-layered brachipatagium in *Rhamphorhynchus* in which the actinofibrils form the dorsal surface, underlain by a layer of fascia, followed by a layer of blood vessel on the ventral surface (Fig. 15C). The composition of the actinofibrils is not known but their external position relative to the wing membrane may give some clue to their identity (Bennett, 2000). The principal structural proteins in vertebrates are collagen and keratin. If the actinofibrils were internal, they would be composed of collagen (as in bats), which allows tensile strength and flexibility. Since they were external, they were probably epidermal structures composed of keratin as in scales and feathers. Bennett argues that keratin is a better material to resist compression and bending than is collagen. Actinofibrils would confer a flatter camber than bat membrane. Such external actinofibrils could have evolved by simple modification of preexisting scales (Padian and Rayner, 1993).

In addition to actinofibrils, well-preserved soft tissues from Brazil reveal that the wing membrane of pterosaur was a complex, multilayered structure, about one mm thick, containing horny epidermis, blood vessels, muscle and thin fibers (Martill and Unwin, 1989). This internal reinforcement and composite

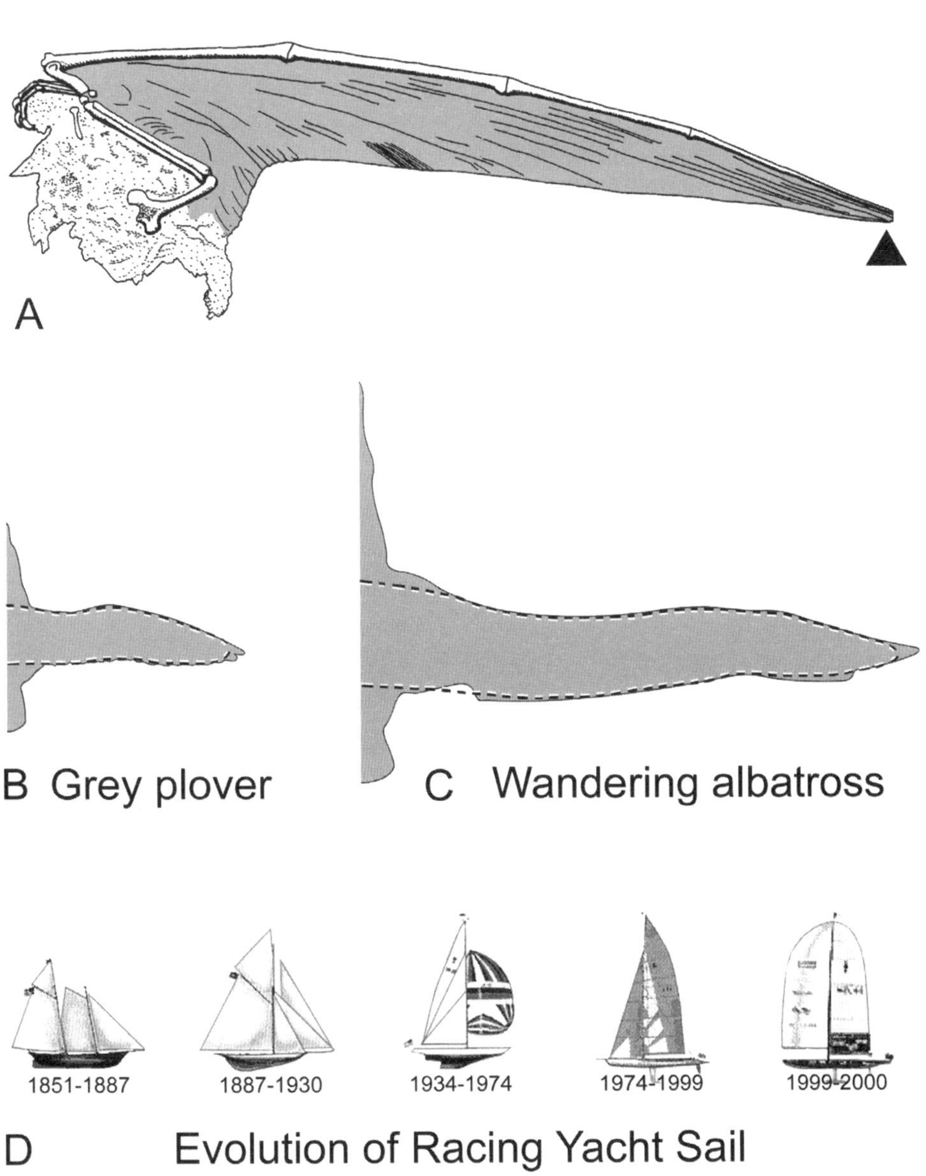

Figure 14. A: The best-preserved pterosaur wing membrane is the *Rhamphorhynchus* wing, described by Zittel (1882), and is referred as the Zittel wing. A simple sketch of the Zittel wing shows a rounded wingtip (modified from Padian and Rayner, 1993). A rounded wingtip is better designed aerodynamically than pointed tip to avoid local flow separation and stalling. B and C: seabirds such as Grey plover and Wandering albatross show rounded wingtips (simplified from Burton, 1990). D: Evolution of the morphology of the sail of a racing yacht at its dorsal edge from sharply pointed to more rounded tip (source: America's Cup Museum, www.acmuseum.com).

nature of the membrane would confer some stiffness in a chordwise direction, preventing it from tearing and billowing during flight. The actinofibrils provide the strength to the membrane to withstand the aerodynamic stresses imposed upon it by its own weight and flapping (Padian, 1985), and also allow folding of the wing in a neat, space-saving fashion like a lady's fan (Fig. 15E), when it was not in use (Bennett, 2000). These fibers run roughly perpendicular to the bones of the arm and hand, but become almost parallel to the wing finger (Fig. 15D). This arrangement of actinofibrils would provide greater bending strength and flexibility to the wing than a flap of skin. The strength is further increased by pinching the fibers together at the knuckle joint to form a fan. This structural reinforcement of the wing may explain why pterodactyloids achieved gigantic size whereas bats failed to do so. It is likely that bat wings would be too delicate to withstand large aerodynamic stresses without structural integrity when scaled to a gigantic size of some pterodactyloids. According to Bennett (2000), the radiating pattern of actinofibrils would prevent the dactylopatagium from narrowing chordwise when it was stretched spanwise laterally. Surprisingly, many insects such as mantids, grasshoppers, and crickets evolved analogous reinforced wings that could be folded in a similar fashion when at rest. Unlike birds and bats, the principal joint of the wing folding mechanism in pterosaurs lies at the knuckle between the wing metacarpal and the first phalanx within the plane of the wing. With the flexion of the knuckle joint, the dactylopatagium could be folded compactly against the wing finger, especially during quadrupedal terrestrial locomotion (Fig. 7). When the wing was

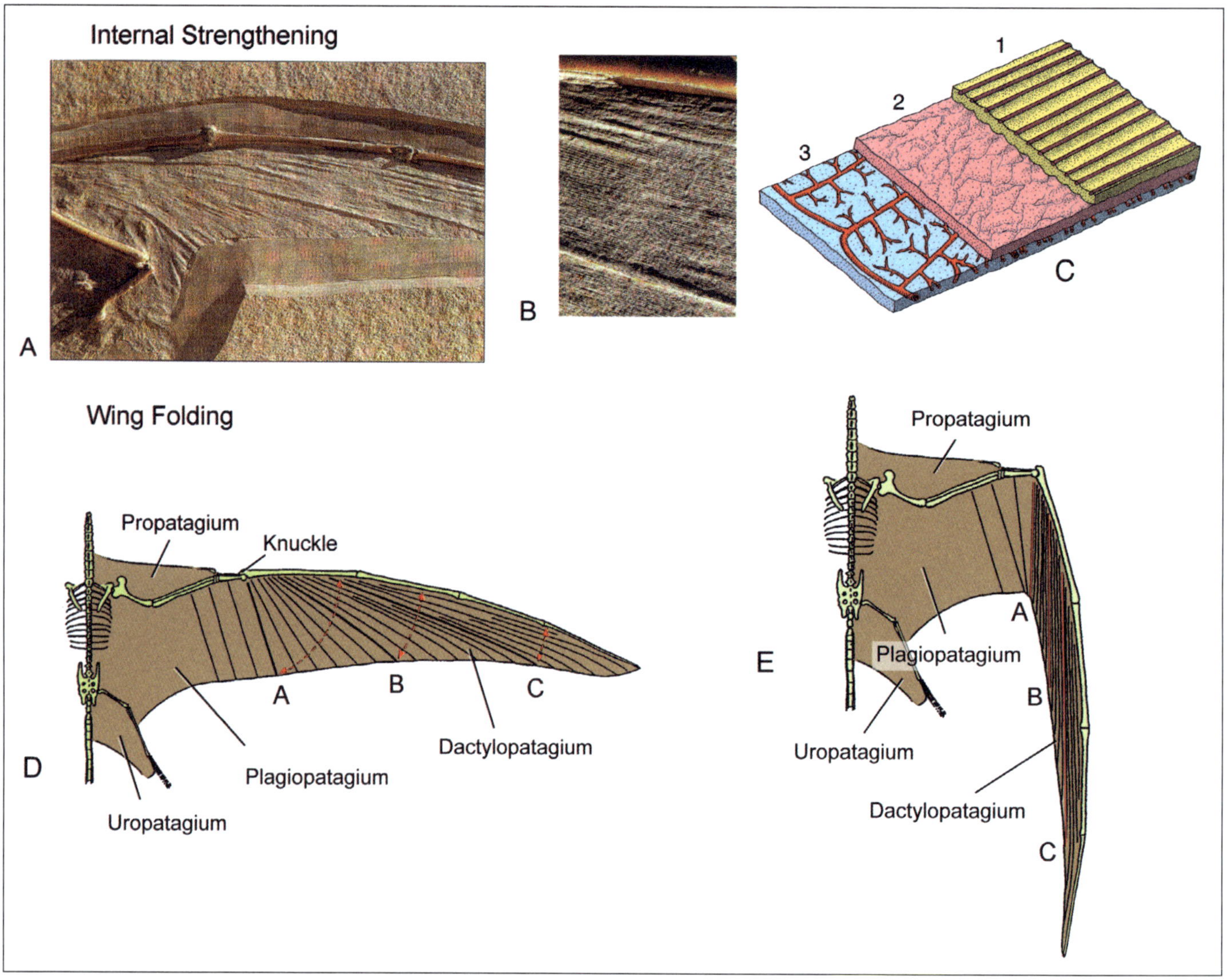

Figure 15. Actinofibrils in pterosaur patagium. A: The Zittel wing of *Rhamphorhynchus* reveals a system of fine parallel fibers, called actinofibrils, especially in the dactylopatagium (after Wellnhofer, 1991a). B: A close-up view of the Zittel wing showing actinofibrils (after Wellnhofer, 1991a). C: Schematic cross section of the wing of *Rhamphorhynchus* showing three distinct layers: 1, actinofibril layer; 2, fascia layer; 3, blood vessel layer (modified from Tischlinger and Frey, 2002). D: Restoration of the Zittel wing in extended position showing the radiating pattern of the selected sets of actinofibrils (after Bennett, 2000). E: The same wing in partially folded position at the knuckle joint showing the role of actinofibrils in compact folding mechanism (after Bennett, 2000).

unfolded by extension of the knuckle joint, these fibers were spread out like a fan (Fig. 15D).

Hair

Superbly preserved specimens of *Sordes* from the Upper Jurassic lake deposits of Kazakhstan reveal that pterosaurs had a dense covering of short hairs up to 6 mm long distributed over most of the head, body, base of the tail, and upper part of the arms and legs (Sharov, 1971; Unwin and Bakhurina, 1994). A coat of hair provided pterosaurs insulation, keeping them warm in winter and cool in summer. The presence of hair in pterosaurs may also indicate the high performance physiology and endothermy demanded by flight. This view is supported by paleohistological study. The great prevalence of fibro-lamellar bone tissues in pterosaurs similar to the conditions seen in modern birds suggests high growth rates, high metabolic levels, altricial birth, and extended parental care (Ricqles et al., 2000). In water spiders, small hairs on the body surface trap air and act as a water repellant. It is likely that pterosaur hair might have functioned in similar way. Since pterosaurs spent a large amount of time feeding in water, a water repellent mechanism would be useful during takeoff to reduce extra weight.

WING ADAPTATIONS FOR POWERED FLIGHT

Skeletal Elements

The skeletons of pterosaurs are epitomes of flight engineering and structure because they combine lightness and strength. Pterosaurs adopted a wide range of flight styles depending on their wing size and body mass. As discussed later, small pterosaurs such as "rhamphorhynchoids" and early pterodactyloids acquired hovering and powered flight and actively flapped their wings. Large pterodactyloids, on the other hand, were mainly soarers but needed flapping strokes during takeoff and landing. The wing joints of pterosaurs are suitable for flapping movements. A framework of skeletal elements that are strongly built to withstand the compressive force of the wing beats supports the wings. Pterosaurs had pneumatic bones to reduce weight as in birds. Unlike bats and birds, the clavicles are absent in pterosaurs. In pterosaurs, both the scapula and the coracoid are powerfully built. They are elongated, L-shaped, and fused together. The glenoid at the confluence of these two bones is a wide, concave facet directed upward, outward, and slightly backward to facilitate the complex movement of the humerus. In basal pterosaurs such as "rhamphorhynchoids," the scapula lies over the dorsal rib cage without any bony connection. However, in Cretaceous pterodactyloids, the upper end of the scapula fits into a socket of the notarium to make the joint immobile. The coracoid is slightly longer than the scapula and is braced strongly against the transverse sulcus of the sternum. The sternal plate is wide and has a shallow ventral keel. Cranial to the coracoidal articulation, the keel continues forward as a narrow protuberance, the cristospine. The cristospine is unique to pterosaurs and is analogous to the manubrium of the bat.

In birds, the scapula, coracoid, and furcula meet at a flexible joint to form the triosseal canal for the passage of the supracoracoideus tendon. In pterosaurs, there is a prominent acrocoracoid process on the coracoid to serve the same pulley function of the supracoracoideus muscle (Padian, 1983). However, because of lack of furcula in pterosaurs, the triosseal canal is open dorsally. In birds, the furcula acts both as a spring and a spacer between the two shoulder girdles (Jenkins et al., 1988). It is also a site for the attachment of the pectoralis muscle. Since the scapula and the coracoid are fused in pterosaurs, and the two bones are braced dorsally and ventrally by the notarium and the sternum, respectively, to form a bony ring, the role of a furcula becomes superfluous for keeping the shoulder girdles apart. Moreover, the cristospine took the role for the attachment of the cranial part of the pectoralis muscle.

Wing Joints

The wing elements of pterosaurs are movable at various joints that permit to change the geometry, camber, and other aerodynamic characteristics of the wing during flapping, gliding, and terrestrial locomotion (Hankin and Watson, 1914; Bramwell and Whitfield, 1974; Padian, 1983; Wellnhofer, 1985, 1991a, 1991b). Some of the joints are mobile while others are rigid. Movable joints occur at the shoulder, elbow, wrist, and knuckle, whereas rigid joints are restricted to the interphalangeal region of the fourth finger. The flexible joints are necessary not only to fold the wing during terrestrial locomotion but also to reduce the span on the upstroke for smaller species. To estimate the range of motion at deformable joints in the wing, we have separated the limb elements from the three-dimensional casts of *Anhanguera* for manipulation.

The wing movement of pterosaurs can be inferred from the motion of the shoulder joint (Fig. 16). Direct articulation of the humerus and the glenoid allows us to estimate the range

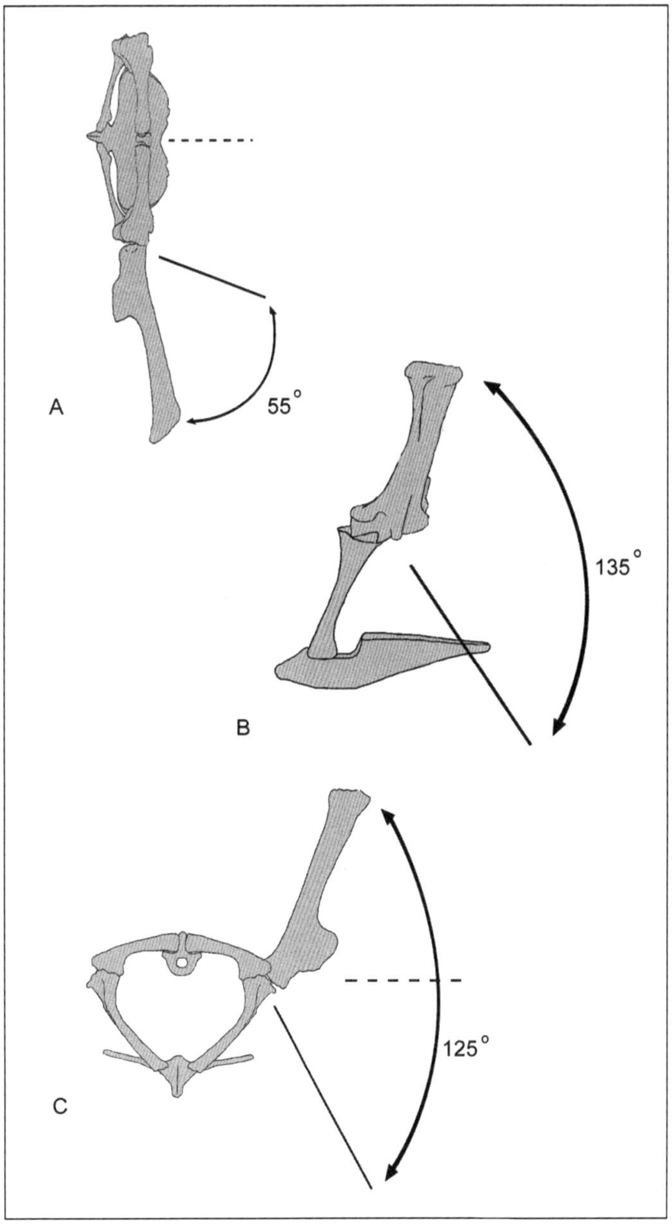

Figure 16. Reconstructed range of motion of the shoulder joint based on manipulations of three-dimensional cast skeleton *Anhanguera piscator* (TTU P10363). A: Dorsal view. B: Left lateral view. C: Cranial view.

of movement of the humerus, and hence the wing. However, reconstructing the motion of joints from fossil specimens is not always straightforward, because they lack cartilaginous capping and epiphyses. The motion at the shoulder joint has been studied previously by several workers: Bramwell and Whitfield (1974), and Bennett (2001) on *Pteranodon*, Padian (1983) and Unwin (1988a) on *Dimorphodon*, and Wellnhofer (1985, 1991a, 1991b) on *Anhanguera* and other pterosaurs. The consensus is that the principal flight motion was primarily alternate depression and elevation of the humerus in a transverse vertical plane, with its distal extremity tracing a sloping oval outline in vertical pathway. In contrast, Hazlehurst and Rayner (1992b) studied the motion of the shoulder joint of *Santanadactylus* and proposed an alternative view. They argued that Cretaceous pterodactyloids were not capable of the wide range of vertical motion of the wing used by birds and bats to maintain horizontal flight. Instead, the shoulder movement was restricted to a 70° rotation of the humerus about its long axis. A long-axis rotation of the humerus is an important component in avian flight. Poore et al. (1997) observed a high-velocity rotation of the humerus of ~70°–80° about its longitudinal axis in pigeons and starlings. The rotation of the humerus, imparted by the supracoracoideus muscle, rapidly elevates the distal wing during the upstroke. Although we concur with Hazlehurst and Rayner that the pterosaur humerus could rotate considerably, we also recognize a wide excursion of its vertical movement. Without vertical movement of the humerus, takeoff and landing would be impossible for pterosaurs.

Our reconstructions are based on manipulations of the shoulder joint of *Anhanguera*. The shoulder girdle is left in its natural position but the wing elements were separated for manipulation. The glenoid is wide with a concave profile vertically, bounded by the lips on the scapula above dorsally and on the coracoid ventrally. The concavity forms an arc in caudal view. The glenoid faces outward, backward, and somewhat upward. The humerus is relatively short, stout with a large deltopectoral crest directed ventrally. The articular head is crescentic in proximal view, is weakly convex, and is continued to its ventral aspect. The shoulder joint permits the humerus to be elevated, depressed, extended, flexed, and rotated about its long axis (Fig. 16). Like birds, one wing beat cycle can be divided into four phases (Jenkins et al., 1988): (1) upstroke-downstroke transition; (2) downstroke; (3) downstroke-upstroke transition; and (4) upstroke. When the articular head of the humerus is placed into the glenoid, the most stable position of the humerus can be inferred. This position may indicate the downstroke-upstroke (or upstroke-downstroke) transition stage during the flight stroke and was extensively used by large pterodactyloids during soaring. In this position, the humerus extends horizontally, making an angle of about 60° to the fore and aft axis in dorsal view; the distal expanded end is oriented almost vertically. From this position, a great range of vertical motion is possible corresponding to the dorso-ventral flapping of the wing. During the upstroke, the humerus is retracted, rotated, and elevated. By late upstroke, the humerus assumes a position of 60° above the horizontal with a rotational component around its long axis. The left humerus rotates clockwise ~40° about its longitudinal axis to supinate the wing. In this position, the angle formed by the long axis of the humerus and the fore and aft axis in dorsal view is ~60°. The upstroke of the humerus is completed and repositions the wing for the next downstroke and assumes the position of upstroke-downstroke transition. During the downstroke, the humerus is extended, pronated, and depressed. By late downstroke, the humerus could be adducted as much as 65° below the horizontal plane, ~125° below the maximum upstroke position. During this position, the long axis of the humerus makes an angle of 80° in relation to the sagittal axis in dorsal view. The rotational component of the humerus around its long axis is ~40° counterclockwise directions to pronate the wing (Fig. 16C).

Our estimate of the range of motion of the shoulder joint (125°) exceeds the range proposed by Wellnhofer (1991a, 1991b). The vertical excursion of the humerus, estimated by other workers, ranges from 95° (Bramwell and Whitfield, 1974) to 90° (Padian, 1983) to 80° (Wellnhofer, 1991a). The total amount of rotation of the humerus on its long axis is about 85°during the upstroke-downstroke cycle, which is comparable to the estimate of Hazlehurst and Rayner (1992b). Padian (1983) suggested that it was possible for the humerus to be folded tightly against the body as in birds, but all other workers considered this position to be impossible; in this position, the head of the humerus becomes entirely separated from the glenoid (Hankin and Watson, 1914; Bramwell and Whitfield, 1974; Wellnhofer, 1991a). Our manipulation of the shoulder joint supports this view. The humerus can be adducted close to the body about 25° in relation to the vertical axis, similar to the range of femoral adduction.

The elbow is a hinge joint enabling a wide range of movement of the forearm in the plane of the wing (Fig. 17). There is some disagreement about the range of excursion permitted at the elbow joint. Padian (1983) using *Dimorphodon* material concluded that the elbow could extend to an angle somewhat less than 180°and flex completely so the wing could be held against the body. However, most workers (Hankin and Watson, 1914; Bramwell and Whitfield, 1974; Wellnhofer, 1991a) considered that the elbow joint was far more restricted, permitting only 30° to 45° closure. Wellnhofer (1991a, 1991b), using material of *Anhanguera* concluded that the elbow could open to an angle of 150° and close to an angle of 110°. We found these values in our specimen as 145° and 90°, respectively. The mechanical constraints at the shoulder and elbow joints clearly indicate that pterosaur wings could not be tightly folded against the body unlike the condition in birds (Figs. 16, 17B).

It is generally believed that the pterosaur wing possesses the ability to synchronize flexion and extension of the elbow and the wrist joints automatically to save muscular effort as seen in birds (Hankin and Watson, 1914; Bramwell and Whitfield, 1974; Wellnhofer, 1991a). This is possible because of a sliding movement of the radius on the ulna similar to a pair of drawing parallels. When the elbow joint is flexed, the radial condyle pushes the radius distally along its axis relative to the ulna. As the radius pushes the leading edge of the radiale, it pushes the distal carpal, which in turn flexes the manus both backward and downward. As a result, the leading edge of the hand undergoes ventral rotation to supinate. In this position, the hand is obliquely oriented to the plane of the wing. Wellnhofer (1991a, 1991b) noticed that in *Anhanguera*, the hands could be angled both backwards and downwards by about 30°. We found a comparable range of motions of the hand at the wrist joint in *Anhanguera*. The wrist joint could open to an angle of 165° and close to an angle of 135° accompanied by a 30° downward component. Birds assume this unusual supinated orientation of the manus during the upstroke when the wingspan is reduced by flexing elbow and wrist joints (Vasquez, 1992). It is likely that pterosaurs would assume a similar pose by flexing knuckle joint during the upstroke to reduce the drag (Fig. 17B).

The principal wing movement in pterosaurs is at the knuckle (metacarpophalangeal) joint, between the fourth metacarpal and the wing finger. Bramwell and Whitfield (1974) estimated a 134°

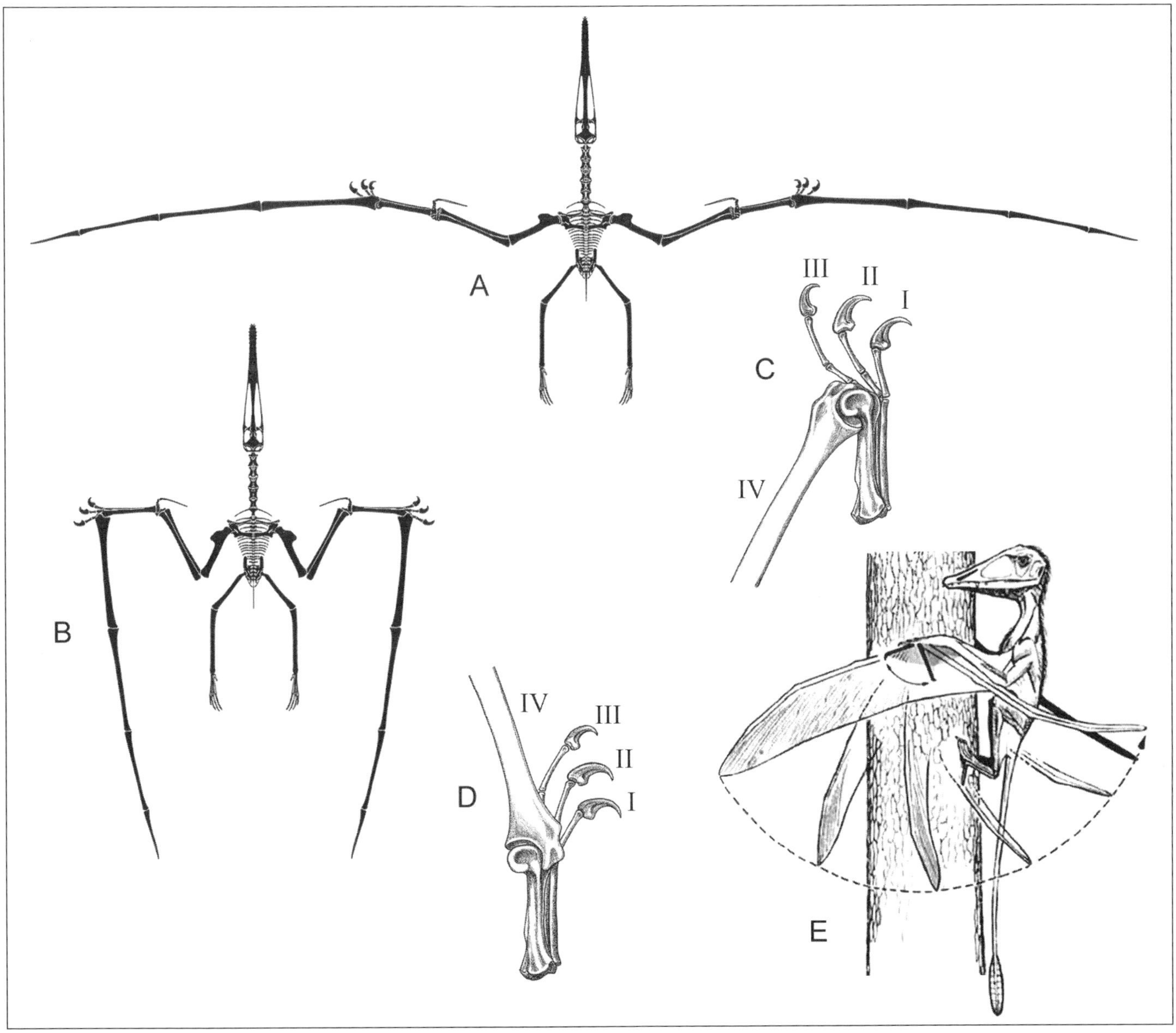

Figure 17. A: *Anhanguera piscator*, skeletal reconstruction on the basis of the cast skeleton TTU P10369; the wings are shown in maximum extension position. B: The wings are shown in maximum flexion position. C: Right wing fingers of *Rhamphorhynchus* showing the opposing folding mechanism at the knuckle joint between the inner fingers (I–III) and the fourth finger (IV). D: The same in extended position; (C and D, simplified from Wellnhofer, 1975). E: Scansorial adaptation of a basal pterosaur, *Preondactylus*, based on Wild (1984) and Peters (2001). Because of the reorientation of the knuckle joint, the fourth finger could be folded or unfolded freely during climbing in an arc, tangential to the perimeter of the tree trunk.

arc of rotation for *Pteranodon*; Wellnhofer (1975) estimated a 140° arc for *Rhamphorhynchus* and 130° arc for *Anhanguera*. We found the knuckle joint could open to an angle of 165°and close to an angle of 35°. However, the flexion and extension of the flight digits is not restricted to the plane of the wing. The condyles at the distal end of the metacarpal are asymmetrical so that when the flight digit is flexed to 35°, it has a small ventral component. This ventral supination of the flight digit is essential for folding of the wing against the body during terrestrial locomotion. The four phalanges of the wing digit are joined by simple concavo-convex joint, which prohibits any motion between them.

The most unusual feature of the pterosaur knuckle joint is the lateral folding mechanism of the fourth finger in the plane of the palm (Fig. 17C, 17D). The manus is highly specialized and differentiated into two components for the dual roles of locomotion. The first three fingers are flexed toward the palm as in other archosaurs and are equipped with terminal claws to aid in quadrupedal walking, when the fourth finger remains passive and folded sidewise. During flight, the first three fingers remain passive, whereas the fourth finger is unfolded to become the flight surface. In this mode, the conjoined metacarpals are oriented in such a fashion that the first three metacarpals of the three small digits are stacked horizontally one behind the other and against the fourth metacarpal. In this position, the first metacarpal forms the leading edge and the claws project downward. The orientation of the metatarsals indicates that the palm is pronated (palmar aspect faces ventrally) and lies in the plane of the forearm. The first three fingers are normally oriented and can be flexed or extended as in other tetrapods. In contrast, the wing finger is movable in the plane of the manus, where the fourth metacarpal is rotated medially along its long axis in such a fashion that its digit can be folded backward. The 90º twist of the fourth metacarpal and associated knuckle joint is a unique feature in pterosaur morphology and is closely linked to the folding and unfolding of the wing. The fourth metacarpal bears an asymmetric hinge joint with the first phalanx of the wing finger that permits the flight digit to be hyperflexed and folded roughly in the plane of the wing. Instead of folding the fourth finger toward the palm at the interphalangeal joint, the whole fourth finger now moves as a unit and can now be folded and extended sidewise in the plane of the palm only at the knuckle joint. The interphalangeal joints are rigid, permitting little or no movement.

Bennett (1997b) argued persuasively for the arboreal leaping origin of pterosaurs. It is likely that the unusual flexion and extension of the wing finger in the plane of the palm may be linked to the climbing arboreal adaptation of protopterosaur on a vertical trunk (Wild, 1984; Peters, 2001). The protopterosaurs acquired various structural adaptations for clinging, hooking, and bracing vertical substrate to avoid falling (Fig. 17E). The claws of both hands and feet were highly recurved similar to those of tree-climbing birds, so they could be dug securely into the bark. They had shorter bodies and longer forelimbs with grasping hands, where the forelimbs were pressed against the lateral side of the trunk, and the body was held in a vertical position. The propatagium and pteroid bone might have acted as an elastic strap to keep the body close to the trunk. The three inner fingers could be used for climbing and clinging to trunks while the fourth finger lacking the terminal claw could be folded sidewise to avoid damage of the membrane. With 90º twist of the fourth metacarpal and knuckle joint, the fourth finger could be flexed and extended freely in an arc, tangential to the circumference of the trunk during climbing. In this lifestyle, the fifth digit, next to the hyper-elongated fourth one became superfluous and was lost in the early history of pterosaurs. The scansorial protopterosaurs developed numerous structural features that were exapted for terrestrial and arboreal locomotion for later pterosaurs. For example, the orientation of the hand during climbing was also used during terrestrial locomotion when the manus was directed laterally and the wing digit was folded sidewise. During flight, the wing digit was unfolded to form a large airfoil surface, while the inner fingers remain passive.

Flight Muscles

Using three-dimensional casts of *Anhanguera*, we reconstructed the flight muscles. Our reconstruction agrees well with previous attempts of restorations of flight muscles in other pterosaurs (Padian, 1983; Wellnhofer, 1991a). In powered flight, two separate motions, the downstroke and upstroke, do most flying. The downstroke provides the forward thrust. The upstroke serves to position the wing for the subsequent downstroke. The osteological features suggest that both sets of muscles responsible for lowering and raising wings are well developed in pterosaurs (Fig. 18).

In birds, both the upstroke and downstroke muscles, pectoralis and supracoracoideus, are anchored to the deep keel of the sternum and insert on the humerus. In pterosaurs, the sternal keel is shallow and needs other areas for the attachment of flight muscles. Moreover, the furcula, another important site for attachment of pectoralis muscle in birds, is absent in pterosaurs. One likely site for the pectoralis is the cristospine, the forward projection of the sternal keel for lowering the humerus. The cristospine in pterosaurs may be analogous for the avian furcula and the manubrium of bats for the attachment of the cranial part of the pectoralis muscle. Because of the shallow keel, the additional site for the attachment of elevator muscles could be the scapula as in bats. Using birds and bats as modern analogs, the flight muscles of pterosaurs are reconstructed (Fig. 19).

In pterosaurs, the large pectoralis muscle, the primary depressor of the wing in downstroke, is anchored to the keel of the sternum and is inserted on the massive deltopectoral crest of the humerus. It has two components: the massive caudal part, attached to the ventral keel of sternum, pulls the wing backward and downward, whereas the cranial branch, which originated at the cristospine, pulls the wing forward and downward. The subdivision of the pectoralis muscle into caudal and cranial components in pterosaurs is interesting and may have some implication during gliding. Many soaring birds such as albatrosses, vultures, and storks lock their wings with subdivided parts of the pecto-

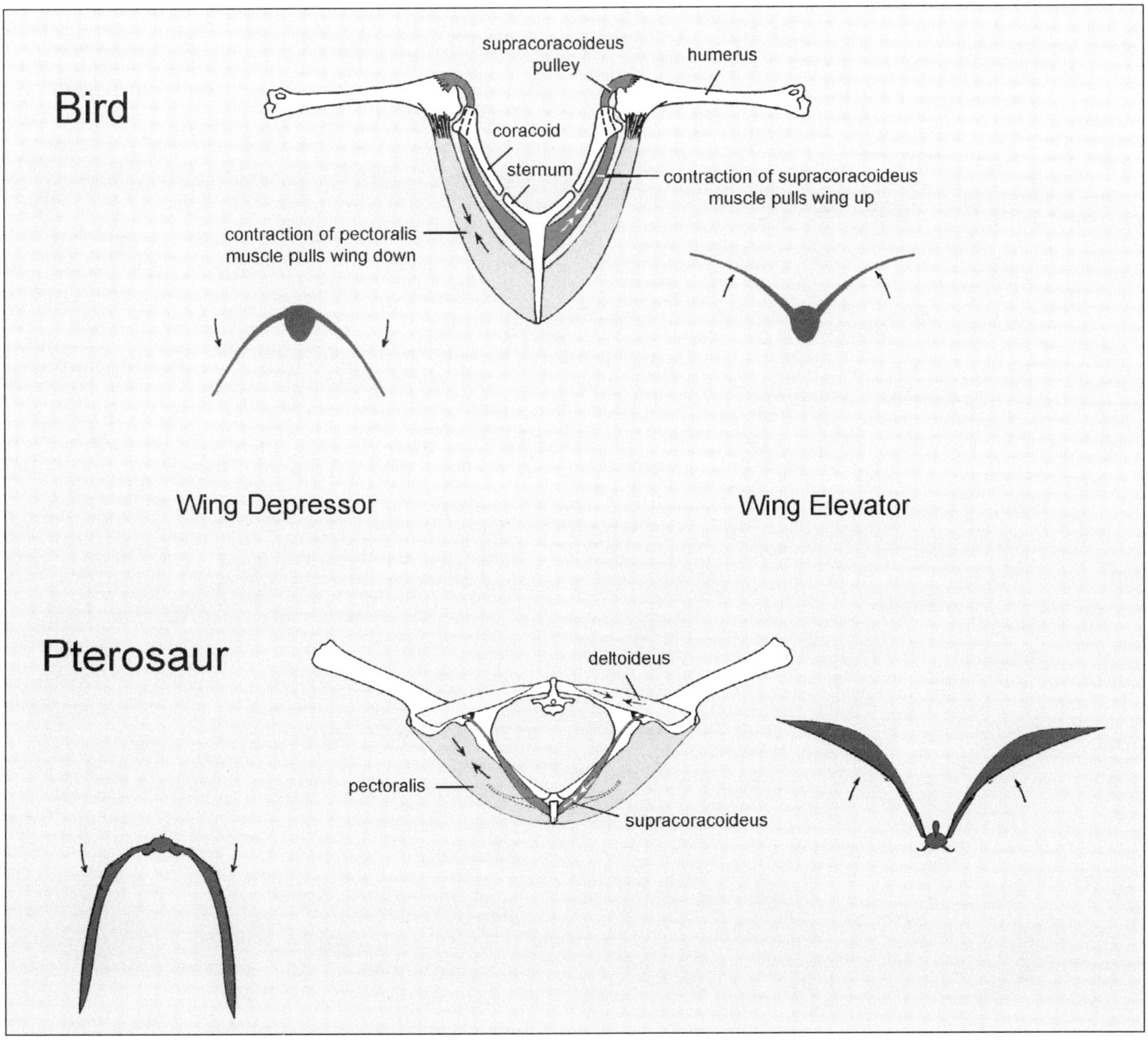

Figure 18. A diagrammatic comparison of the flight muscles of bird and pterosaur in cranial view showing how they operate during flapping. In pterosaurs, the supracoracoideus, the elevator muscle, was weak relative to birds as is evident from the shallow keel of the sternum, and was supplemented by another elevator muscle, deltoideus.

ralis muscle preventing the wing from being elevated above the horizontal (Pennycuick, 1982). In these birds, the smaller, deeper part of the pectoralis is a slow tonic muscle that holds the wings in place when gliding. It is reasonable that the caudal part of the pectoralis in pterosaurs might have functioned in a similar way to lock the wing during gliding. Moreover, in albatrosses, there is an additional locking device of the wing during soaring. A fan-shaped ligament locks the shoulder joints when the wings are outstretched and fully spread for soaring. A similar ligamental connection could have been present in pterosaurs for locking the shoulder joint during soaring.

The upstroke movement is orchestrated by three groups of muscles inserted on various points on the dorsal surface of the humerus. The supracoracoideus muscle lies deep to the pectoralis along the sternal keel. It passes dorsally as a tendon through the acrocoracoid process and inserts on the proximal dorsal surface of the humerus to operate like a pulley. Since the sternal keel is shallow in pterosaurs, the supracoracoideus muscle was weak in pterosaurs, relative to birds. Two additional elevator muscles compensated this weakness: the deltoideus and latissimus dorsi. The deltoideus muscle originating from the proximal part of the scapula rotates the humerus outward and upward, thus pulling the wing up and to the rear. The latissimus dorsi originates from the supraspinous ligament of the thoracic spine and inserts on the proximal surface of the humerus. It adducts and flexes the wing, moving it backward dorsally.

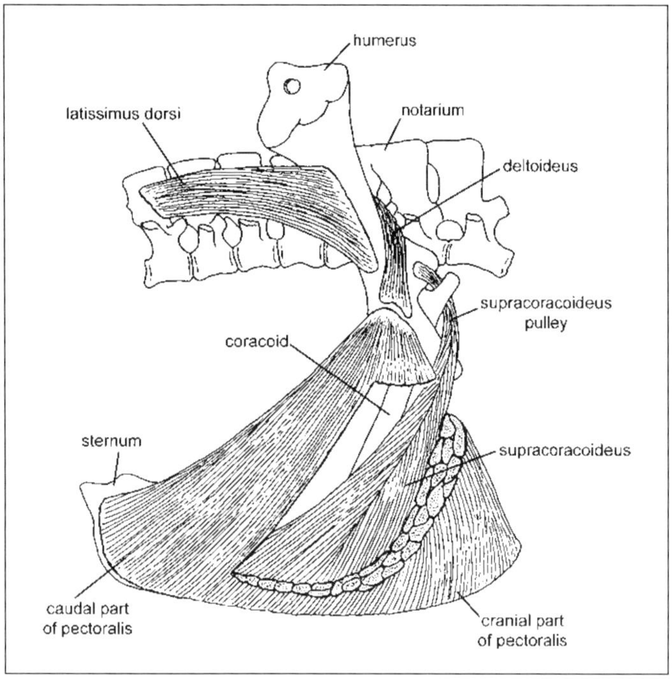

Figure 19. Reconstruction of the flight muscles in *Anhanguera* in right lateral view. The principal depressor muscle, the pectoralis is anchored to the keel of the sternum and pulls the humerus down. The cranial part of the pectoralis pulls the wing forward and downward, whereas the caudal part may lock the wing in horizontal position during soaring. The elevator muscles, which pull the humerus up are located at different locations. The supracoracoideus lies underneath the pectoralis muscle (shown by cutaway section) and is attached to the sternal keel. Dorsally, it forms a pulley near the shoulder joint and is inserted on the humerus. Another branch of the elevator muscle is deltoid, which appears to be well developed in pterosaurs. The third branch is the latissimus dorsi pulls the wing backward and upward.

Stability and Neural Control

Pterosaurs with long, narrow wings had a high aspect ratio and a high lift/drag ratio that would have contributed significant lift-based stability during flight. Flight requires fine coordination between muscles and senses. The neurological basis of control holds particular interest in that flight demands on sensory integration, equilibrium, and muscular coordination are acute. The brain sends signals to nerves, which operate muscles. The flying animal demands acute vision for guidance and a large cerebellum for balance and coordination. The development of stable flapping flight required the integration by the cerebellum of the continuous proprioceptive input from the muscles with the output from the motor centers. Recent high resolution CT scans of pterosaur braincase reveal cerebral and cerebellar expansion, ventrolateral displacement of the optic lobes, great enlargement of semicircular canals and floccular lobes (Witmer et al., 2003). The most striking aspect of pterosaur brain morphology is the space devoted to the vestibular system and the sense of equilibrium in the cerebellar area. The enlarged flocculus in pterosaurs may relate to receiving proprioceptive afferents from the wing. Tischlinger and Frey (2002) had documented that the wings of pterosaurs had a ventral layer of blood vessels, muscles, and tendons, which would have carried muscle spindle fibers back to the central nervous system (Fig. 15C). The wing might have acted as a big sensory organ, where the enlarged flocculus might have acted as an autopilot device to process proprioceptive and other somatosensory information from the wing that stretched between the limbs, as well as with the limb joint themselves, thus having a more direct impact on flight control (Witmer et al., 2003).

When a pterosaur's flight was disturbed by turbulence, the movement was detected by the highly enlarged semicircular canals of the inner ear; compensations were then made by fine adjustments of the wings, and normal flight was resumed. The large flocculus projecting sidewise from the semicircular canal would perceive pressure disturbances as well as change of flow patterns over the wings and would act as an autopilot device for balance and control. Pterosaurs could respond to constantly varying conditions using their sensors to monitor pressure variations over the entire surface of the wing. They could morph their wing shape to exploit flight conditions and control speed.

AERODYNAMIC CONSTRAINTS

Pterosaurs had a 160 million year history of aerial locomotion and adapted to different styles of flight. Although Cuvier (1801) recognized the flying ability of pterosaurs two centuries ago, fascination in the flight of pterosaurs was rekindled at the beginning of the twentieth century with the invention of the airplane. Many aeronautical engineers, zoologists, and paleontologists became intrigued to unravel the mystery of the pterosaur flight. In most aerodynamic analyses, *Pteranodon*, the large, crested pterodactyl from the Niobrara Formation of Kansas with a wingspan of 7 m, became the model (Hankin and Watson, 1914; Kripp, 1943; Heptonstall, 1971; Bramwell and Whitfield, 1974; Stein, 1975; Brower, 1983). The consensus was that *Pteranodon*, with its long, narrow wing, was primarily a long-distance soarer similar to extant seabirds such as frigatebirds and albatrosses. In addition to the theoretical approaches of pterosaur flight, several aeronautical engineers such as Holst (1957), Winkworth (1985), and MacCready (1985) actually built flying models of pterosaurs to simulate their flapping, gliding and soaring capabilities.

We selected ten species of pterosaurs with a wide spectrum of size, from the smallest to the largest forms, to study their flight performance (Chatterjee and Templin, 2001). Some basic aerodynamic data are needed to analyze the flight characteristics of pterosaurs. Some parameters, such as body length (distance from the first dorsal vertebra to the last sacral vertebra) and wingspan, can be measured directly from the skeletons or skeletal restoration. An accurate reconstruction of the outstretched wing in dorsal aspect is required to measure the wingspan. These and other parameters, such as mass, wing area, wingspan, aspect ratio and wing loading, are used for analysis of the flight performance of pterosaurs (Table 3). The wingspan of selected species ranges from 0.4 m to 10.4 m, and the mass from 0.015 kg to 70 kg. Thus the largest pterosaur in our study weighs about 4700 times more than the smallest species, and the longest wingspan is 25 times the shortest (Fig. 20).

Mass

Various methodologies have been developed for estimating body mass in fossil species. The most common approach is the use of a scaling relation between lengths or diameters of limb elements with mass based on univariate regression.

Jerison (1973) used this simple allometric equation to estimate the mass of pterosaurs:

$$W = 0.5L^2 \tag{1}$$

where W is body weight (in g), and L is the head and body length (in cm).

Yalden (1984) used the following equation to estimate the mass of *Archaeopteryx*:

$$W = \left(\frac{l}{36}\right)^{-0.28} \tag{2}$$

$$M = 13.25d^{2.353} \tag{3}$$

where W is the weight (kg), M is the mass (g), l is the length of femur (cm), and d is the diameter of the femur (mm).

Templin (2000) suggested a simple equation between body length l (m) and mass M (kg) as follows:

$$l = \left(\frac{M}{35}\right)^{\frac{1}{3}} \tag{4}$$

All the methods described above (Equations 1–4) are univariate predictive regressions where the body mass is regressed against some univariate measure of body size (limb length, limb diameter of body length). For all data sets, multivariate predictions of body mass were more accurate than univariate predictions (Atanassov and Strauss, 2002). Brower and Veinus (1981) used multivariate analysis and provided various allometric equations derived from geometric models of the pterosaurs, where mass (M in g) can be estimated as a function of wing area (S in cm^2).

$$M = 0.0408\,(S)^{1.27} \tag{5}$$

They also provided the relationships of wing loading (WL in g/cm^2) as) as a function of mass (M in g).

$$WL = 0.07472(M)^{0.227} \tag{6}$$

TABLE 3. BASIC AERODYNAMIC DATA FOR 10 SPECIES OF PTEROSAURS

Species	Body length l (m)	Mass M (kg)	Wing area S (m^2)	Wing span b (m)	Aspect ratio AR= b^2/S	Wing span/ body length Ratio b/l	Wing loading WL= Mg/S (N/m^2)	Root-chord/ body length ratio c_o/l	Power Available P_{avai} (W)	Cruising speed m/sec
1. *Eudimorphodon ranzii*	0.075	0.015	0.021	0.412	8.08	5.49	7.00	0.865	0.608	3.8
2. *Pterodactylus antiquus*	0.108	0.0386	0.0274	0.538	10.56	5.23	13.60	0.631	1.13	5
3. *Rhamphorhynchus muensteri*	0.156	0.134	0.072	0.864	10.34	5.52	18.25	0.678	2.62	6.4
4. *Dorygnathus banthensis*	0.188	0.232	0.104	0.920	8.14	4.89	21.88	0.766	3.776	8
5. *Tapejara wellnhoferi*	0.229	0.418	0.166	1.35	10.98	5.91	24.70	0.684	5.59	8
6. *Nyctosaurus gracilis*	0.376	1.86	0.409	2.72	18.08	7.23	44.60	0.509	15.12	9.6
7. *Dsungaripterus weii*	0.512	4.70	0.747	3.24	14.05	6.33	61.70	0.573	28.10	13
8. *Anhanguera piscator*	0.700	7.58	2.118	4.69	10.39	7.81	35.10	0.958	38.59	11.6
9. *Pteranodon longiceps*	0.780	16.6	2.650	6.95	18.22	8.91	61.40	0.623	65.10	13
10. *Quetzalcoatlus northropi*	1.760	70.0	9.55	10.39	11.30	8.25	71.88	0929	170.00	16

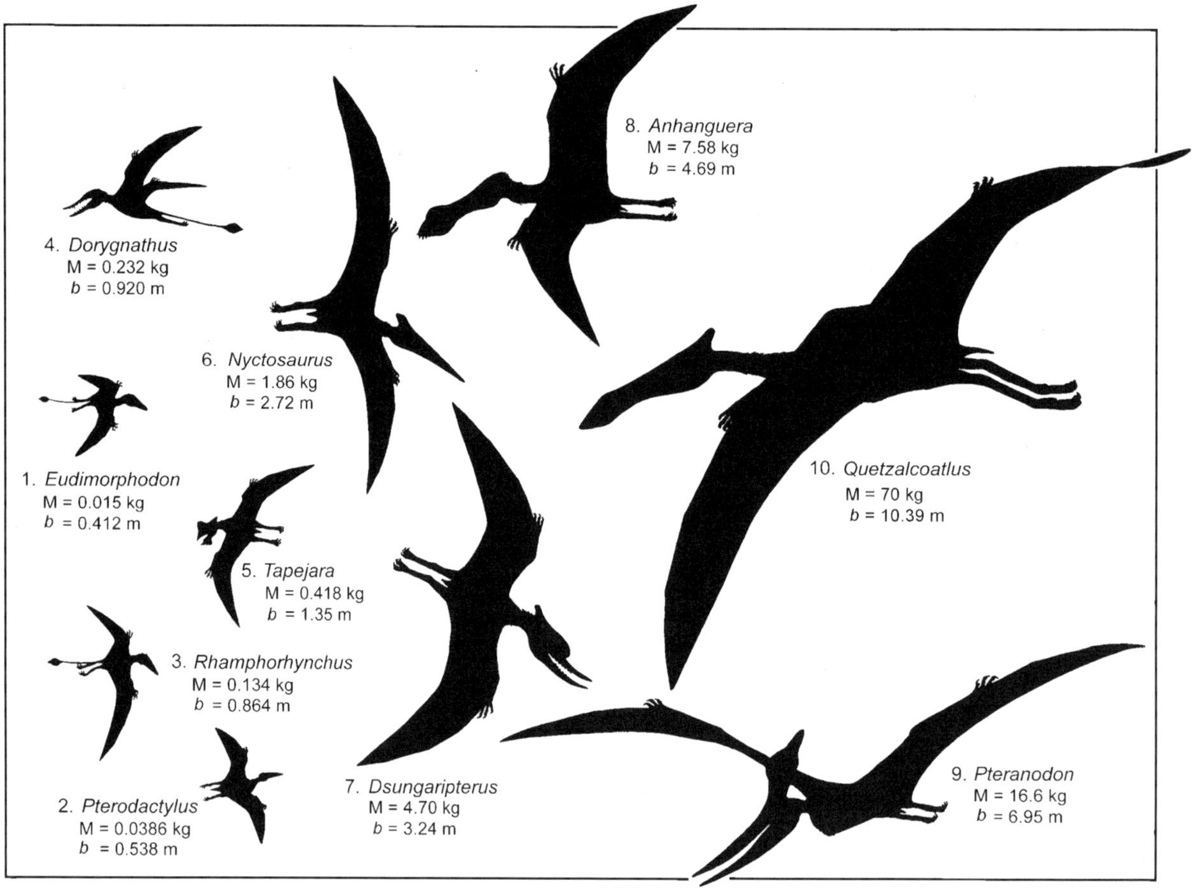

Figure 20. Size range of ten genera of pterosaurs used in this study with their wingspan (*b*) and mass (*M*). The body mass ranges from 0.015 kg to 70 kg and the wingspan from 0.4 m to 10.4 m. Thus the largest pterosaur in our study weighs about 4700 times more than the smallest species, and the longest wingspan is 25 times the shortest.

A better method for predicting body mass is to calculate the volume from geometric modeling of the animal and to multiply it by its density (~0.73g/cm^3 for modern birds) (Bramwell and Whitfield, 1974; Brower and Veinus, 1981; Hazlehurst, 1991; Hazlehurst and Rayner, 1992a). Body measurements (15 characters) and estimated volume of six species of pterosaurs—*Eudimorphodon ranzii*, *Rhamphorhynhus muensteri*, *Dorygnathus banthensis*, *Nyctosaurus gracilis*, *Dsungaripterus weii*, and *Pteranodon longiceps*—were taken from Brower (1980) and Hazlehurst (1991) and were used to calibrate the regression. The same set of body measurements were taken for *Tapejara*, *Anhanguera* and *Quetzalcoatlus* directly from casts at the Museum of Texas Tech University. For *Pterodactylus antiquus*, measurements were taken from Wellnhofer (1970). Using multiple regression analysis, body masses of *Pterodactylus*, *Tapejara*, *Anhanguera* and *Quetzalcoatlus* were estimated (M.N. Atanassov and R. Strauss, 2002, personal commun.).

Previous mass estimates of *Quetzalcoatlus* are problematical, ranging from 200 kg (Paul, 2002) to 85 kg (Brower and Veinus, 1981). Initially, we used both 200 kg and 85 kg mass estimates to calculate the flight performance of *Quetzalcoatlus*. It is unlikely that a 200 kg *Quetzalcoatlus* could take off from the ground or fly, as revealed from our initial flight simulator study (contra to Paul, 2002). Similarly, we found that it would be difficult for an 85 kg *Quetzalcoatlus* to take off with a run, even with full anaerobic power and headwind, as well as to maintain powered flight. Based on the two density estimates, a simple multivariate method (principal-component analysis) suggests that the body mass of *Quetzalcoatlus* ranges from 62 to 77 kg (M.N. Atanassov and R. Strauss, 2002, personal commun.). In our final analysis, we used a mass of 70 kg for *Quetzalcoatlus* (midway in the range given by principal components regression) to study its flight capability. A 70 kg mass for *Quetzalcoatlus* appears to be optimum limit for gliding, takeoff, and landing, assuming it was well adapted in the Big Bend ecology. All calculations were performed using the program *Matlab*® version 5. Our estimated masses for the following six species of pterosaurs match well with the published accounts: *Eudimorphodon* and *Dorygnathus* (Hazlehurst and Rayner, 1992a); *Nyctosaurus*, *Dsungaripterus*, and *Pteranodon* (Brower and Veinus, 1981).

For the remaining five taxa—*Rhamphorhynchus*, *Pterodactylus*, *Tapejara*, *Anhanguera*, and *Quetzalcoatlus*, we provide the estimated masses (Table 1).

Wing Area and Wingspan

The estimation of the wing area (symbol S) is complicated by the uncertainty regarding the caudal extension of the patagium to the leg. The wing planform was reconstructed in two morphotypes, wide wing for "rhamphorhynchoids" and narrow wing for pterodactyloids (Fig. 12). We estimated the wing area from a restoration of the outstretched wing planforms in dorsal view by digitizing the body outline using the program tpsDig® version 1.31 by Rohlf (2001). Wingspan (symbol b) is the transverse width from one wingtip to the other, with the wings spread, and the chord is the distance from the front to the rear edge of the wing.

Aspect Ratio

Aspect ratio (*AR*) is a familiar term in aerodynamics. It measures the slenderness of a wing, the ratio of tip-to-tip wingspan (b) to average width (S/b), where S is the total wing area; so $AR = b^2/S$. It is a dimensionless index of the shape of the wing, being high for a long narrow wing, and low for a short broad one. Theoretically, the coefficient of induced drag, which is the penalty that must be paid for lift production, is proportional to the square of the lift coefficient and inversely proportional to the aspect ratio. As will be explained below, perhaps to the surprise of some readers, when induced drag and lift coefficients are converted to actual induced drag and lift (with newtons as the units of force in the international standard SI metric system) the wing area and aspect ratio disappear, leaving induced drag proportional to the square of the so-called span loading (L/b), where L is the aerodynamic lift, with average value equal to animal weight (Mg) in level flight. Nevertheless, aspect ratio and wing area or wing loading (Mg/S) are important parameters in other respects. For instance the minimum drag-to-lift ratio and minimum slope in gliding flight are inversely proportional to $AR^{1/2}$. Seabirds with long, narrow wings include terns, gulls, frigatebirds, and albatrosses, all of which have aspect ratios ranging from 10 to 19. As a general rule, the higher the aspect ratio the better the aerodynamic performance of a wing. In the ten pterosaurs studied, the aspect ratios similarly range from 8 to 18 (Fig. 21). As an illustration of the different effects of aspect ratio and wingspan, we may compare the gliding behavior of the golden eagle *Aquila chrysaetus* and the giant petrel *Macronectus giganteus*, using data from the Appendix of Tennekes (1996). Both have roughly the same mass (4.8 and 5.3 kg, respectively) and the same wingspan (2.0 m). But their aspect ratios are very different (6.1 and 12.0, respectively). Owing to their similar wingspans, they have about the same induced drag at each speed and very nearly the same minimum sinking speed in glides, but the eagle is able to glide and circle more slowly at around 10 m/s with its larger wing area and lower wing loading, whereas the petrel has a flatter glide angle at ~15–20 m/s.

Wing Loading

Wing loading ($WL = Mg/S$, with units N/m^2) is another important parameter that affects the performance of flying animals. Flying speed in flapping and gliding flight is approximately proportional to the square root of the wing loading. Larger animals (and aircraft) have higher wing loading and faster speeds than do small ones. Small pterosaurs with low wing loadings are more maneuverable than are large pterosaurs, and the larger animals have to fly faster to stay airborne but can glide very efficiently in windy environments, such as the open ocean.

The wing loading of ten species of pterosaurs ranges from 7 N/m^2 in *Eudimorphodon* to 72 N/m^2 in *Quetzalcoatlus* (Fig. 21). Tennekes (1996) provided basic aerodynamic data of 14 species of seabirds (Table 4). Since the wing design of pterosaurs is remarkably similar to that of seabirds, we plotted the weights against wing loadings of ten species of pterosaurs and 14 species of seabirds. We used a log-log scale because of the large mass range, and because a power function or allometric Equations would plot as a straight line with slope equal to the exponent (Fig. 22). For each group, there is a distinct regression line indicating a close relationship between weight and wing loading. The straight line that fits the pterosaur data shows a slope of 0.27 whereas for bird data, the slope is 0.40. We also plotted weights against cruising speeds in Figure 23, using the data for seabirds listed by Tennekes (1996), and the cruising speeds calculated for the pterosaurs in this study. Again, each group shows a distinct regression line. The straight line that best fits the pterosaur data has a slope of 0.17 whereas for the bird data, the slope is 0.20. The largest seabirds have significantly larger wing loadings and higher cruising speeds than pterosaurs of the same weight. This may be an indication of their different lifestyles: the seabirds are fast open-sea gliders and the large pterodactyloids perhaps were efficient soarers, as in the comparison above between the giant petrel and golden eagle.

Weight

Weight is another important factor in flying performance. Weight (W) is often confused with mass (M). Weight (Mg) is the product of mass (M) and gravitational acceleration ($g = 9.81$ m/s^2) and is often expressed in newtons (N). From the estimated mass, we have calculated the weight of ten pterosaur species (Table 5).

Aerodynamic Performance Estimation

When an aircraft is flying horizontally, its weight is balanced by the average lift on the wings, and the drag on the wings and body is balanced by the forward thrust. In a shallow glide in still air, there is no propulsive thrust, but there is a component of gravity acting forward along the sloping flight path. Unlike airplanes, flying vertebrates get both lift and thrust from their wings, and many are capable of hovering flight at zero airspeed. In this respect, they are like helicopters, which depend on their rotors for both support and thrust. In this study we have used helicopter

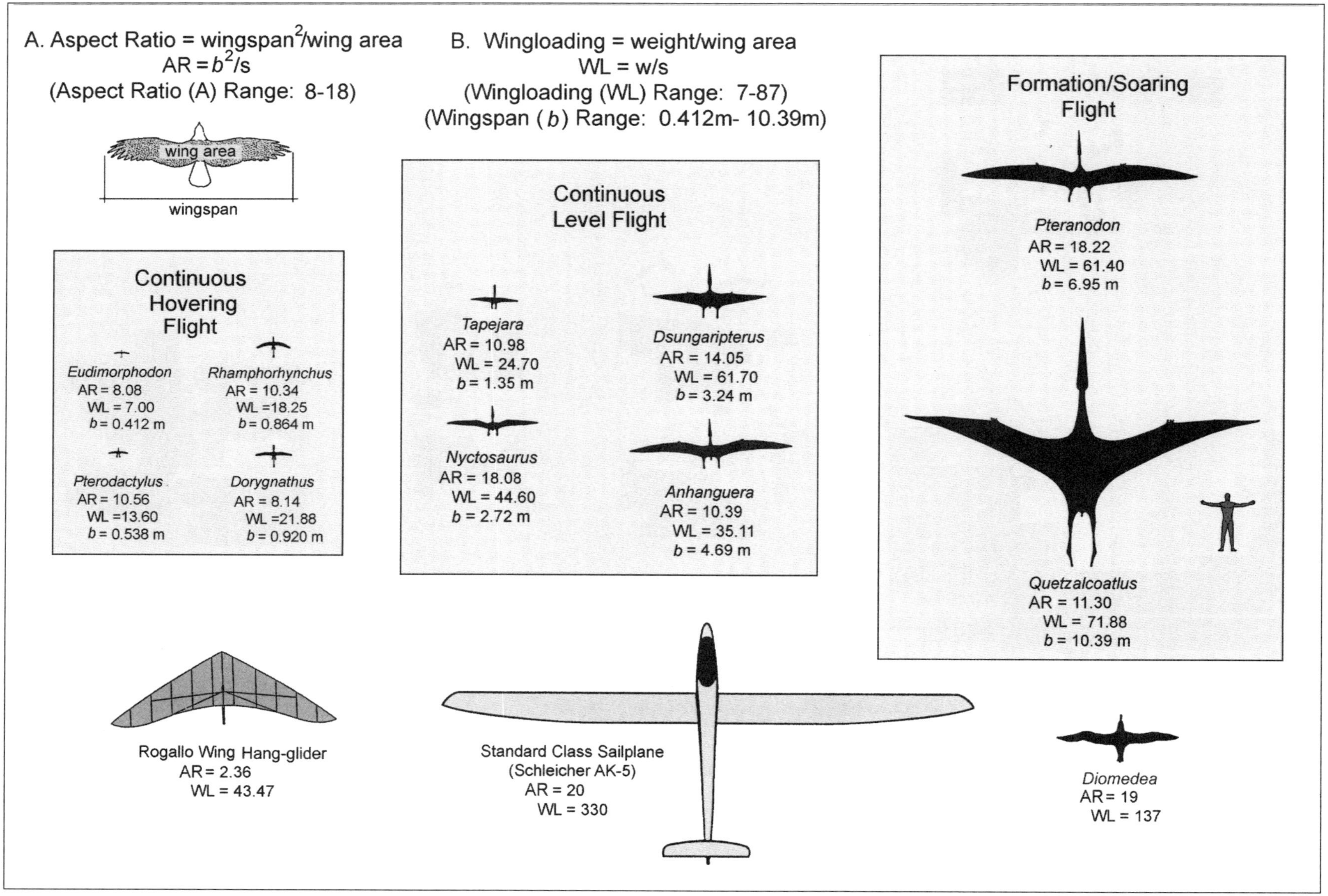

Figure 21. Comparative morphologic and aerodynamic data for ten genera of pterosaurs and other flyers such as hang glider, sailplane, and albatross *Diomedea*. The flight styles of pterosaurs can be grouped into three categories on the basis of size and other aerodynamic data: small pterosaurs for hovering flight, medium-sized pterosaurs for continuous level flight, and the large pterosaurs for formation/soaring flights. Data for hang glider (after Brower, 1983) and sailplane and *Diomedea* (after Tennekes, 1996). Large pterosaurs adapted similar soaring style like modern large seabirds such as frigatebirds and albatrosses.

TABLE 4. WEIGHT, WING AREA, WING LOADING, AND AIRSPEEDS FOR VARIOUS SEABIRDS (COMPILED FROM BROWER, 1983, AND TENNEKES, 1996)

	Weight W (N)	Wing Area S (m^2)	Wing Loading W/S	Velocity m/sec	Velocity mph
1. Common tern	1.15	0.050	23	7.8	18
2. Dove prion	1.70	0.046	37	9.9	22
3. Black-headed gull	2.30	0.075	31	9.0	20
4. Black skimmer	3.00	0.089	34	9.4	21
5. Common gull	3.67	0.115	32	9.2	21
6. Kittiwake	3.90	0.101	39	10.1	23
7. Royal tern	4.70	0.108	44	10.7	24
8. Fulmar	8.20	0.124	66	13.2	30
9. Herring gull	9.40	0.181	52	11.7	26
10. Great skua	13.50	0.214	63	12.9	29
11. Frigate bird	16.20	0.324	50	11.4	26
12. Great black-backed gull	19.20	0.272	71	13.6	31
13. Gannet	26.90	0.245	110	17.0	38
14. Sooty albatross	28.00	0.340	82	14.7	33
15. Black-browed albatross	38.00	0.360	106	16.7	38
16. Giant petrel	42.55	0.285	150	19.9	45
17. Wandering albatross	87.00	0.620	140	19.2	43

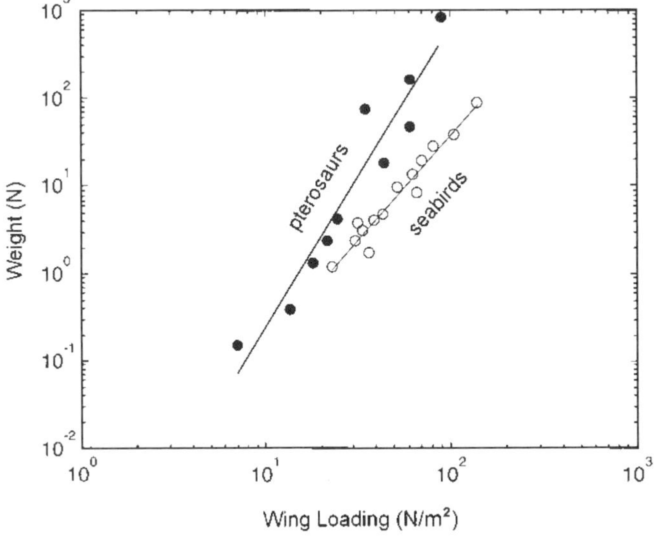

Figure 22. Weight for pterosaurs and seabirds, when plotted against wing loading on logarithmic coordinates, tends to fall along two distinct straight lines. In both cases, the wing loading increases with the weight of the animal. The slope of the regression line for pterosaur data is 0.269, whereas for seabird data, the slope is 0.397. Pterosaurs had slightly lower wing loading compared to seabirds of the same weight. Data for pterosaurs (see Table 3) and seabirds (Table 4).

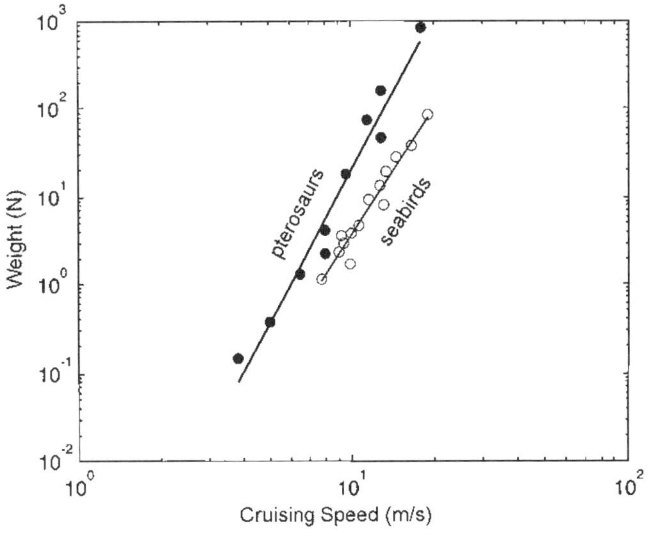

Figure 23. Weight for pterosaurs and seabirds, when plotted against cruising speed on logarithmic coordinates, tends to fall along two distinct straight lines. In both cases, the cruising speed increases with the weight of the animal. The slope of the regression line for pterosaur data is 0.168, whereas for seabird data, the slope is 0.198. Pterosaurs had slightly lower cruising speed compared to seabirds of the same weight. Data for pterosaurs (see Table 3) and seabirds (Table 4).

TABLE 5. BASIC TERRESTRIAL TAKEOFF DATA FOR 10 SPECIES OF PTEROSAURS

Species	Mass M (kg)	Weight W (N)	Wind speed (m/s)	Ground slope (deg)	Running speed (m/s)	Power (W)	Liftoff distance (m)	Liftoff speed (m/s)
1. *Eudimorphodon ranzii*	0.015	0.15	0	0	1.4	0.61	0.19	0
			5	0		0.61	0	
2. *Pterodactylus antiquus*	0.039	0.38	0	0	1.9	1.14	0.30	0
			5	0		1.14	0	
3. *Rhamphorhynchus muensteri*	0.136	1.33	0	0	2.2	2.62	0.56	0
			5	0	2.2	2.62	0	
4. *Dorygnathus banthensis*	0.232	2.28	0	0	3	3.776	0.48	3.4
			5	0	3	3.776	0	3
5. *Tapejara wellnhoferi*	0.418	4.10	0	0	3	5.59	1.20	3.8
			5	0	3	5.59	0	3
6. *Nyctosaurus gracilis*	1.86	18.25	0	0	3.4	15.12	4.86	4.76
			5	0	3.4	15.12	0	3.5
7. *Dsungaripterus weii*	4.70	46.11	0	0	4.7	28.06	33.56	6.77
			5	0	4.7	28.06	0	5
8. *Anhanguera piscator*	7.58	74.36	0	0	4.8	33.06	0	5
			5	0	4.8	33.06		
9. *Pteranodon longiceps*	16.6	162.84	0	0	5.6	100*	18.8	7.27
			5	0	5.6	100*	0	6
10. *Quetzalcoatlus northropi*	70.0	686.5	0	0	10	500*	196	14.94
			5	0	10	500*	0	10
			0	–10	10	170	21.37	13.04
			5	–10	10	170	0	10
			0	–10	10	0	no takeoff	
			5	–10	10	0	0	10

*Anaerobic power

momentum stream tube theory, which may be unfamiliar to most readers, to analyze the flight performance of pterosaurs. A full account of the method as applied to flying animals is given in Templin (2000), but since that reference in the aeronautical literature, may not be readily available, and in any case is rather technical, a summary is given here.

Most functional analyses of flight performances of flying animals have centered on birds, taking advantage of aerodynamic equations used in aircraft design. The most common method for approximate performance estimation is an aircraft-like model (Prandtl and Tietjens, 1934), which uses familiar parameters such as lift and drag coefficients, wing area, wing aspect ratio, and span efficiency. In this method, the power required to maintain steady level flight is calculated as the product of aerodynamic drag and flight airspeed. The total drag is calculated as the sum of two components: the so-called induced drag, which is the penalty that must be paid for the production of aerodynamic lift, and a component (sometimes called zero-lift drag) assumed independent of lift, comprising pressure drag and surface skin friction. This simple method has a limitation: it predicts infinite induced drag and power at zero airspeed and therefore cannot be used without modification for performance estimation in hovering or near-hovering flight. A second model, known as momentum streamtube theory, was originally proposed for straight wings by Ludwig Prandtl (Prandtl and Tietjens, 1934), developed further by helicopter designers (Stepniewski and Keys, 1984), and adapted to animal flight (Templin, 2000). It avoids the zero-speed problem by making the assumption that a cylindrical tube of air having a cross-section area (A) with diameter approximately equal to the wingspan (b), or rotor diameter, initially approaching the wing or rotor at flight speed (V), is deflected downward through an angle (θ), which may vary from 90° in hovering at zero speed to a small angle in cruising and high-speed flight. The flight power is equated to the change in kinetic energy flow in the streamtube from far upstream to far downstream, which is required to balance lift and aerodynamic drag. In comparison with the aircraft method, the streamtube model has its own drawbacks: unfamiliarity to most animal flight researchers, and greater numerical complexity in application. However, it is important to note that the two methods produce essentially identical results in unpowered (gliding) and in powered flight at speeds that are usually above 5–10 m/s.

The streamtube model is not new. The principles behind the diversity in flying behavior among birds are the same principles Ludwig Prandtl discussed more than 70 years ago and are more suited to study animal flight than is the aerodynamics of the 1980s (McMahon and Bonner, 1983). By the early 1920s, Prandtl (see Prandtl and Tietjens, 1934) had developed the so-called lifting line wing theory, in which a wing of span b is replaced by a spanwise "bound" vortex extending from tip to tip, with a trailing vorticity sheet, strongest near the wingtips, leaving the bound vortex in the downstream direction. He showed that the lift-induced drag was a minimum when the lift distribution along the span was elliptic in shape, shedding constant downwash velocity across the span in the wake. In that case the induced drag coefficient is:

$$C_{Di} = \frac{C_L^2}{\pi AR} \tag{7}$$

This is the familiar induced drag coefficient relation, and in practice the denominator is usually written πeAR, for better agreement with experimental measurements, the parameter e being the so-called span efficiency, typically of the order of 0.8–0.9. Sometimes its reciprocal $k = 1/e$ (~1.1–1.25) appears instead in the numerator. In unpowered gliding flight, we set $e = 0.85$, but allow for its possible reduction in flapping-wing flight.

If the definitions $C_{Di} = D_i/1/2\rho V^2S$, $C_L = L/1/2\rho V^2S$ and $AR = b^2/S$, are substituted into Prandtl's Equation, the induced drag, including the efficiency e, is obtained as:

$$D_i = \left(\frac{2}{\pi e\rho V^2}\right)\times\left(\frac{L}{b}\right)^2 \tag{8}$$

The symbol definitions are D_i = induced drag (N), L = lift (N), ρ = air density (1.225 kg/m^3 in the standard atmosphere at sea level), V = airspeed (m/s), b = wingspan (m) and S = wing area (m^2). This will explain the fact, already noted, that the induced drag (as distinct from its aerodynamic coefficient) is not explicitly a function of either wing aspect ratio or wing area, but does depend on the span loading L/b (N/m).

A common method used by other workers (e.g., Pennycuick, 1972; Bramwell and Whitfield, 1974; Brower, 1983) for the estimation of drag in steady flight is to assume that it is the sum of three components, for example:

Total drag D = wing profile drag + body drag + induced drag =

$$\left(\frac{1}{2}\rho V^2\right)\left(C_{DW}S + C_{Dbody}S_{body}\right) + \left[\frac{2\left(\frac{Mg}{b}\right)^2}{\left(\pi e\rho V^2\right)}\right] \tag{9}$$

In this equation, the component drag coefficients C_{DW} and C_{Dbody} are each multiplied by their respective reference areas. An alternative method, which we use in this study, is to estimate an effective skin friction coefficient C_F for the wing and body separately, using the "wetted" surface area of for each component rather than by S and S_{body} as in Equation 9. The wetted area of the wing is $2S$, and that of the body is estimated to be approximately equal to l^2, where l is the body "formula" length given by Equation 4. This area is somewhat larger than the actual surface area of an ellipsoidal body of length l and diameter about 30% of l, to make an allowance for enlargement due to other elements such as feathers, fur, head, legs, etc. Owing to their relatively small bodies, the body drag of pterosaurs is usually much smaller than the wing profile drag. The effective friction coefficient is a function of the Reynolds number Re, which is a dimensionless measure of aerodynamic scale effects, proportional to the ratio of inertia forces to viscous forces in the moving fluid. It is defined as $Re = \rho VL/\mu$, where L (in this case) is a representative length in the flow direction, usually the body length or the mean width (chord length) of the wing, and μ is the fluid viscosity. Usually Re is numerically large; when the values of air density and viscosity at sea level and temperature are inserted, $Re = 69{,}000 \times VL$ (V and L units are m/s and m). For a large pterodactyl at, say 10 m/s speed with a mean wing chord of 1 m, the wing Re is ~700,000, but of course is much lower for the smallest pterosaurs. Based on empirical data from several sources (see Templin, 2000, fig. 17), we use the following relation for the skin friction coefficient:

$$C_F = \frac{2.7}{\sqrt{Re}},\ \text{but } C_F = 0.005 \text{ for } Re > 300{,}000 \text{ approximately} \tag{10}$$

The relation at low Re, based on wind tunnel and oil channel measurements on airfoils and bodies, is twice the theoretical coefficient for fully laminar (non-turbulent) flow, decreasing steadily with increasing Re. As Re increases beyond the above critical value, the coefficient levels out, as the fluid boundary layer along the surface becomes fully turbulent. The fixed value (0.005) is derived from the upper range of measured drag of streamlined bodies (e.g., Hoerner, 1992) and also agrees with the zero-lift drag coefficient $C_{do} \approx 0.01$ for sand-roughened airfoil sections at high Re (Abbott and Doenhoff, 1959). Our wing profile drag coefficient, based on wing area S is therefore never less than 0.01. This method is clearly based on aircraft-like surfaces conditions, though not perfectly "smooth"; a collection of drag measurements on non–jet fighter aircraft of 1940s vintage (Perkins and Hage, 1949) suggests that the "equivalent" overall skin friction coefficient ranged from about 0.003 to 0.006.

The last term in Equation 9 is the induced drag according to Equation 8, with lift L replaced in level flight by the weight Mg. The power P (in watts, W) required to balance the total drag in steady flight is the product of total drag (N) and speed V (m/s), with (in some cases) an increment added to allow for the difference between fixed-wing and flapping-wing flight. Obviously, Equation 9 cannot be used for the estimation of induced drag of induced power during hovering or in slow flight because of the quantity V^2 in the denominator, so some other method must be found to fill in the low speed power gap. This problem does not occur in the streamtube method, which has no lower speed limit.

In the development of his lifting line vortex theory, Prandtl also showed (see for example Prandtl and Tietjens, 1934, p. 110) that the induced drag relation (Equation 7) can also be derived if the simplifying assumption is made that a cylindrical tube of fluid diameter b—and cross-section area $A = (\pi/4)b^2$—initially approaching the wing at flight speed V, is deflected downward through a small angle in reaction to wing lift. Taking into account of the span efficiency factor the effective streamtube area becomes:

$$A = \left(\frac{\pi}{4}\right)eb^2 \tag{11}$$

Later, the streamtube model was modified by helicopter engineers and expanded to include the effect of propulsive power input to the system, and to remove limitations on the value of the downwash angle and on the fluid speeds along the axis of the tube (Stepniewski and Keys, 1984). We further adapted the model to flapping wing–wing flight on the assumption that it refers to the *average* values of lift, drag, and power throughout the flapping cycle. A schematic sketch of this idealized flow model is shown in Figure 24.

The pterosaur is immersed in the stream tube, which has a diameter equal to $\sqrt{e}$ times the wingspan b. Three values of velocity relative to the animal are shown: the flight velocity V, the velocity V_1 through the disk area A, and a final wake velocity V_2, which is deflected through angle θ relative to the flight direction. In unpowered flight, all three velocities are equal in magnitude, but in powered flight V_1 and V_2 are greater than V owing to the addition of propulsive energy. The mass flow in the streamtube is $\rho V_1 A$. In vector notation (bold type) the relation between the three velocities is $\mathbf{V_1 - V = 1/2(V_2 - V)}$; i.e., the vertical and horizontal components of induced velocity at the actuator disk are one-half those far downstream. This condition was found by Prandtl for the downwash velocities behind the wing lifting line and was extended to the helicopter rotor by Stepniewski (Stepniewski and Keys, 1984, vol. 1, p. 62). The aerodynamic lift L (equal to the animal weight Mg in level flight or nearly so in shallow glides) is equated to the vertical component of the change in the momentum vectors, $\rho V_1 A(V_2 - V)$, and in level powered flight, the net aerodynamic forward thrust is equated to the sum of the wing profile drag and body drag, which are calculated as in the first two terms on the right-hand side of Equation 9. When these conditions are taken into account, the initially unknown speeds and wake deflection angle can be found as functions of flight speed, air density, and animal parameters from the momentum vector geometry.

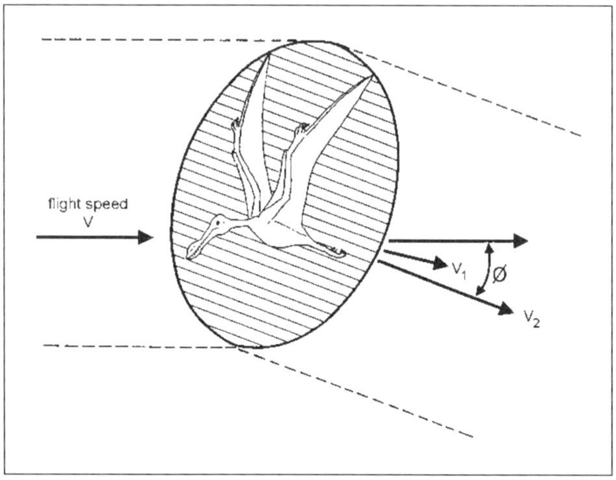

Figure 24. Momentum stream tube vectors showing the diagrammatic sketch of the imaginary circular stream tube that is "captured" and deflected as it approaches and passes through the animal's wing system. The approach speed is the flight speed V. The tube is deflected through an angle of θ as lift is produced.

Finally, the power required for equilibrium flight is given by the increase in kinetic energy flow in the streamtube from the approaching speed V to the final speed V_2:

$$P = \frac{1}{2}(\text{mass flow rate})\left(V_2^2 - V^2\right) = \frac{1}{2}(\rho V_1 A)\left(V_2^2 - V^2\right) \tag{12}$$

A summary of the method of estimating the power required at slow flight speeds is as follows. The momentum theory gives the power required to hover at $V = 0$) as:

$$P_H = \left[\frac{(\text{Mg})^3}{2\rho A}\right]^{\frac{1}{2}}, \text{ where } A = \left(\frac{\pi}{4}\right)eb^2 \tag{13}$$

At very low flight speeds, the total power required for steady level flight consists almost entirely of so-called induced power P_i since the remainder, the zero-lift power as approximately proportional to V^3 and can be neglected at low speed. In that case, P_i at any airspeed V can be determined from equation (12) if the magnitudes of the other two velocities V_1 and V_2 shown in Figure 24, are known. They can be determined from the solution of two algebraic equations derived from the geometry of the streamtube momentum vector diagram. The momentum flow rates through the actuator area A (defined in Equation 11), and in the downstream wake flow are given respectively by:

$$\rho V_1^2 A = \left[(\rho V V_1 A)^2 + \left(\frac{\text{Mg}}{2}\right)^2\right]^{\frac{1}{2}} \tag{14}$$

and

$$\rho V_1 V_2 A = \left[(\rho V V_1 A) + (\text{Mg})^2\right]^{\frac{1}{2}} \tag{15}$$

in which the product Mg is the animal weight. Further algebra can be simplified, at least in appearance, by defining a reference velocity V_{ref}:

$$V_{ref} = \left(\frac{2\text{Mg}}{\rho \text{A}}\right)^{\frac{1}{2}} \tag{16}$$

For example, for a *Pteranodon* for which the air density $\rho = 1.225$ kg/m^3 (sea-level standard density), $M = 16.6$ kg, $b = 6.9$ m, $e = 0.7$ near hovering, from which $A = 26.2$ m^2, $V_{ref} = 3.18$ m/s. We now redefine the three velocity vectors in non-dimensional form as

$v = V/V_{ref}$, $v_1 = V_1/V_{ref}$ and $v_2 = V_2/V_{ref}$. When the velocities in Equations 14 and 15 are divided by V_{ref}, they become, respectively:

$$v_1^2 = \left[(vv_1)^2 + \left(\frac{1}{4}\right)^2\right]^{\frac{1}{2}} \text{ and } v_1 v_2 = \left[(vv_1)^2 + \left(\frac{1}{2}\right)^2\right]^{\frac{1}{2}}$$

When squared on both sides, they become quadratic equations, with solutions:

$$v_1^2 = \frac{1}{2}\left[v^2 + \left(v^4 + \frac{1}{4}\right)^{\frac{1}{2}}\right] \text{ and } v_2^2 = v^2 = v^2 + \left(\frac{1}{2}v_1\right)^2$$

Hence also

$$v_2^2 - v^2 = \left[\frac{1}{(2v_1)^2}\right] \quad (17)$$

Using Equations 12, 17, and the definition of V_{ref} to change V's to v's, the induced power P_i becomes:

$$P_i = \left(\frac{1}{4v_1}\right)\left[\frac{2(\mathrm{Mg})^3}{\rho A}\right]^{\frac{1}{2}} \quad (18)$$

If this is divided by the value of the hovering power P_H, given by Equation 13, the ratio of P_i/P_H is:

$$\frac{P_i}{P_H} = \left\{2\left[v^2 + \left(v^4 + \frac{1}{4}\right)^{\frac{1}{2}}\right]\right\}^{-\frac{1}{2}} \quad (19)$$

The hovering power P_H is given by Equation 13, and induced power P_i is obtained from Equation 19 after the non-dimensional speed v is calculated from $v = V/V_{ref}$. Total level flight power P is the sum of P_i and zero-lift power, the latter being the product of speed V and the body and wing zero-lift drag, which in this study has been estimated by the method described previously, using Equations 9 and 10.

A few values of the induced power ratio and actual induced power for *Pteranodon*, as functions of dimensionless speed and actual flight speed V (m/s) are as follows:

Dimensionless speed v	0	0.2	0.4	0.6	0.8	1.0	1.2	1.4
Pteranodon speed V m/s	0	0.64	1.27	1.91	2.54	3.18	3.82	4.45
P_i/P_H from Equation 19	1.0	0.96	0.85	0.72	0.59	0.49	0.41	0.35
Pteranodon power (W)	259	249	220	186	153	127	106	91

When the dimensionless speed is much greater than 1.0, Equation 14 reduced to $P_i/P_H \approx 1/(2v)$, approximately. It can be shown that this is identical with the induced power that is obtained by multiplying the induced drag term Equation 9 by the speed V. Thus for $v > 1$ (or in the case of *Pteranodon*, for $V > 3$ m/s, approximately) the streamtube model gives the same induced drag and power as the more conventional drag estimation method.

For powered flight, it is also necessary to have an estimate of the maximum steady (aerobic) mechanical power available for wing flapping. Maximum steady metabolic rates for birds were obtained from several sources such as Tucker (1970, 1973) and Peters (1983, table IIId). After allowing for a conversion efficiency of 20% (Pennycuick, 1972; Tucker, 1975) an empirical fit to the data was found to be (see Templin, 2000, fig. 13):

$$P_{avail}(W) = 10\left[M(\mathrm{kg})\right]^{\frac{2}{3}} \quad (20)$$

We use two computer programs, named ANFLTPWR and ANFLTSIM, for streamtube performance equations. ANFLTPWR computes the power required and available in equilibrium level flight at all speeds and ANFLTSIM simulates time dependent takeoffs, landings and flight paths in a vertical plane. Allowance is made for variation of the span efficiency (e) as a function of a certain power coefficient based on power, speed and wingspan, since there is evidence (e.g., Tucker, 1973; Dial et al., 1997) that the apparent value e decreases to a minimum value (e_{min}) during hovering at $V = 0$ and high power, but increases to a maximum (e_{max}) at high speed and in unpowered gliding. We usually set e_{max} = 0.85 and e_{min} = 0.5 to 0.7 at $V = 0$ for birds. This version of the theory was originally designed to applicable to the smallest insects as well as vertebrate flyers, and for small insects e may be as low as 0.02 or 0.03 (Bartholomew and Casey, 1978). The allowance for variable e makes computer iteration necessary, since e is not known precisely at each step until next approximation to power is computed. However, convergence is fast, usually requiring 2 to 3 steps. In ANFLTSIM, this iteration is unnecessary since the power (P), which may be varied during flight, corresponding to throttle control in aircraft, is one of the "pilot's" inputs. Other variable inputs are the wake angle deflection θ, corresponding to attitude and lift variation through movement of a control column in an aircraft, and "aerobrake" actuation (e.g., by extension of legs or wing pitchup) is also included. In this program also, iterations are sometimes necessary, since a running check is kept on the approximate average lift coefficient of the flapping wings, and either power or lift are automatically reduced to keep within a specified value of C_{Lmax}. ANFLTSIM also includes a pitch damping subroutine, with adjustable gain, which continuously adjusts the lift around the input value of θ to decrease or eliminate the long-period pitching undulations that occur in powered or gliding flight if the pitch control is "frozen."

Some general remarks must be made regarding the reliability of the method. Rayner (1979) has criticized it, and by implication the more conventional vortex field method summarized in Equation 9 as well, since it oversimplifies the real flow field, particularly its unsteadiness. Rayner's criticism is valid, at least in principle, but his vortex field method has its own difficulties, including computational complexity and the need for detailed knowledge relating to wing morphology and wing flapping geometry, which have to be known or estimated in each case.

Figure 25 is an updated version of Figure 2 in Templin (2000). Its coordinates are mass M (kg) and the span/length ratio b/l, where l is a "formula" body length, an empirical function of mass as given by Equation 4 above. For reasons already mentioned, this ratio is preferred to the wing aspect ratio AR as an indication of performance capability, though they are closely related. Different flying animals such as birds, bats, and pterosaurs occupy their own distinct areas in the chart, each showing the range of their flying styles as size increases.

The four shaded mass limits were calculated using the streamtube model. The limit for safe gliding was based on many ANFLTSIM glides launched at 20 m height and ending in rapid high-drag pitchups reaching final speeds not greater than 5 m/s. North American flying squirrels, the Philippine colugo, and several species of small Australian gliding marsupials (not shown) extend to the left along the "safe gliding" limit as far as $M = 0.12$ kg and $b/l = 0.6$, and the largest flying bird, the extinct *Argentavis magnificens*, is apparently also a safe glider at about $M = 70$ kg and $b/l = 5$. Similarly, the other limiting lines, computed using ANFLTPWR, lie roughly where they would be expected, relative to the regions occupied by large and small flyers. For example, the migrating Canada goose (*Branta canadensis*) lies above the mass limit for continuous level flight but below that for formation flight. The small pterosaurs were

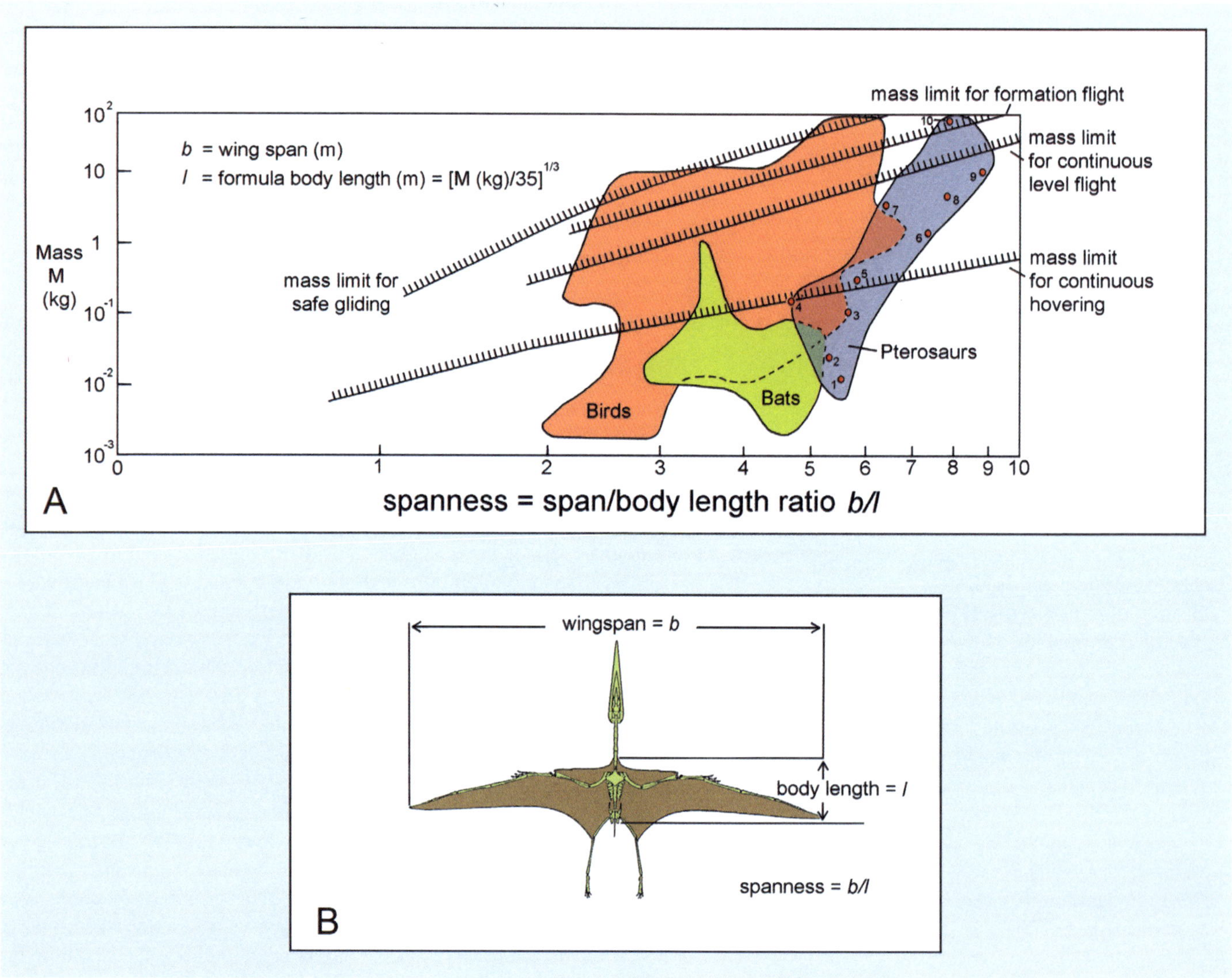

Figure 25. A: Relation between mass and spanness of three groups of flyers—birds, bats, and pterosaurs. The chart shows four sloping hatched bands, the lower edges of which correspond to the theoretical estimates in the upper mass limits, respectively: (1) for continuous hovering flight; (2) for continuous level flight; (3) for formation flight; and (4) for safe gliding. The smaller pterosaurs share with similar-sized bats and birds in hovering flight. The medium and large-sized pterosaurs overlap with similar-sized birds with flapping and gliding flights. B: Ventral view of a pterosaur showing two important parameters, wingspan (b) and body length (l).

apparently capable of hovering flight, and some birds, such as large short-winged fowl, are incapable of any kind of continuous flight. Agreement between the predicted and radar-measured cruising speeds of several species migrating birds (Burton, 1990) have been excellent for all except the three smallest of 14 species. In general, the results of the method, when applied across a wide spectrum, have been highly satisfactory.

It is important to point out, however, that the method as it is used in this study gives lower estimates of total drag and level flight power for pterosaurs than those proposed by some other authors (see for example Brower's [1983] power curves for *Nyctosaurus* and *Pteranodon* in our Figure 29)]. Since it was shown above that the streamtube model gives very nearly the same induced drag as the conventional drag estimation method except at slow flight speeds, total drag differences must be attributed to differences in the assumed wing profile or body drag coefficients, rather than to the model itself. Bramwell and Whitfield (1974) in their aerodynamic analysis of *Pteranodon*, and Brower (1983) in his drag analysis of *Pteranodon* and *Nyctosaurus* used considerably higher values of the wing profile drag coefficient, based on measurements with thin, highly cambered wings, such as have been used in hang gliders or ultralight aircraft. As already mentioned, our minimum wing profile drag coefficient decreases as the Reynolds number increases, but for the largest pterosaurs at high *Re* (high flight speeds) we have $C_{\mathrm{Dprofile}} = 0.01$ at zero lift, whereas Bramwell and Whitfield estimated this coefficient to have a minimum value of 0.017 at $C_L = 0.8$ to 0.9, rising to 0.036 at low C_L. Brower's corresponding values are about twice as high as Bramwell and Whitfield's at similar lift coefficients. We believe that the explanation for the high drag coefficients at low lift coefficients (i.e., at high flight speeds) is flow separation that develops behind the wing leading edge on the *lower* surface at low angles of attack, a kind of negative stall leading to a thick turbulent underwing wake, and that it is more likely that a streamlined fairing behind the spar and an ability to flatten the airfoil camber at high speed may have evolved (see Fig. 13, A–D), since the performance benefits from modest changes in geometry would have been large.

GLIDING AND SOARING

Gliding

Gliding is a simpler and cheaper way of flying than flapping wings. It is energetically economical and a relatively fast mode of travel. Gliders stretch their motionless wings to form one lifting surface and passively descend through the air by the aid of gravity. The combined results of two forces, downward pull of gravity and upward and forward lift move the glider ahead. The forward component of the gravity is sufficient to overcome the drag on the body. The higher the lift-to-drag ratio the glider has, the better it will glide. Gliding is a passive mechanism of flight where height is continuously lost. Good gliders cover long distances with little descent. However, when the rate at which an upward wind current lifts a glider is equal or greater than its sinking speed, it can stay aloft indefinitely.

Hazlehurst and Rayner (1992) compared the shapes of pterosaur wings with those of birds through the use of Principal Component Analysis (PCA). Plotted onto a loading-aspect ratio graph, all pterosaurs, regardless of wing planform reconstruction, reside within the low loading-high aspect ratio quadrant. This quadrant is also occupied by the extant marine soarers. Therefore, all pterosaurs examined had low flight speeds, high maneuverability, and good soaring performance, which are attributed to low loading values. The high aspect values indicate high levels of aerodynamic efficiency.

Much of our understanding of the gliding flight of birds comes from the pioneer work of Pennycuick (1968). Pennycuick used a tilted wind tunnel to study the gliding skill of birds. He trained a bird to fly in the jet of air. Because of the inclined position of the artificial jet, the bird did not need to flap but could glide into the wind. The wings remained stationary so that the observer could measure the aerodynamic data. In this setup, Pennycuick studied the gliding ability of the bird by varying the speed of the jet and the angle of the tilt. He recorded the shallowest angle at which the bird could glide.

Gliding performance is measured by a polar curve, which is a plot of horizontal speed (V) versus sinking speed (w). Sinking speed is determined primarily by the wing loading, so that lighter, larger gliders descend more slowly than do smaller, heavier ones. Sinking speed is equal to the ratio between the power (P) needed to maintain horizontal flight and the weight (W), and is called power loading. This can be expressed as a simple equation (Tennekes, 1996):

$$w = \frac{P}{W} \tag{21}$$

Previous work upon the flight performance of the pterosaurs has often been based upon the construction of glide polars (Heptonstall, 1971; Bramwell and Whitfield, 1974; Brower, 1983). Pterosaurs with long wings of high aspect ratio had a shallower glide. A glide polar is expressed by fitting a curve to real or calculated data. It shows three critical points. The sharp downturn on the left side of the polar curve indicates the minimum airspeed at which a glider can fly, or its stall speed. The point of minimum sinking speed is at the top of the curve where the flyer can glide for the longest duration from a given altitude. The third point represents the airspeed at the best gliding ratio. At this point, pterosaurs could glide the longest distance from a given altitude. The speed for the best glide angle (maximum lift/drag ratio) can be found by drawing a tangent between the glide polar and the origin. Flying animals usually glide at speeds faster than the minimum sinking speed to avoid stalling. When a glider glides slowly, it loses height quickly because the induced drag is high. Similarly, when it glides fast, it also loses height quickly, because the profile drag is high.

Gliding performance of pterosaurs can be estimated by plotting the horizontal speed against the vertical or sinking speed. We used the ANFLTSIM program to calculate these values. We plotted the glide polar curves for ten species of pterosaurs (Fig. 26). We find discrepancies in gliding performance between our curves and those estimated by Brower (1983). This was due to high wing profile drag coefficients and low lift coefficients of the thin, highly cambered airfoil that was used by Brower. Since the drag coefficients increase as the lift coefficient decreases (opposite to what is common in more conventional airfoils), a possible boundary layer separation on the lower (ventral) surface behind the leading edge, which becomes more severe as the angle of attack is reduced. As noted earlier, pterosaurs might have solved this problem by streamlining and stretching the wing to flatten the wing camber.

The glide polar curves show that there is a regular progression of gliding performance in pterosaurs as size increases. Larger pterosaurs have better gliding performance. Pterosaurs have minimum sinking speeds around 0.4 m/s in the present analysis, largely regardless of size. However, the airspeed for the best gliding angle does depend on size, increasing with increasing mass and wing loading, from as low as 4 m/s for *Pterodactylus* to 14 m/s for *Quetzalcoatlus*. Thus, the best glide angles of large pterosaurs are less than 2° downward; they are nearly equivalent to modern sailplanes, but better than hang gliders, which have good lift but high drag. Small pterosaurs have best lower gliding airspeeds, so their best gliding angles are 4° or more. The minimum sinking speed and the best gliding ratio are excellent indicating that pterosaurs were skilful gliders (Fig. 26). Note that the glide curve for the wandering albatross is similar to that for *Quetzalcoatlus*, although they are very different in mass.

Gliding performance is often calculated as a lift/drag (L/D) ratio, called the glide ratio (Fig. 27A). French engineers aptly named the glide ratio as finesse (F) (Tennekes, 1996). We have plotted ten polar curves of pterosaurs in log-log coordinates and compared them with those of albatross, pheasant, ultralight, sailplane, Fokker Friendship, and Boeing 747 (Fig. 27B). If a pterosaur were to stay aloft effortlessly for an extended period, its rate of descent should be equal or less than 1 m/sec. This is the arbitrary soaring limit. It appears from Figure 27B that most pterosaurs were excellent soarers, like modern seabirds, with

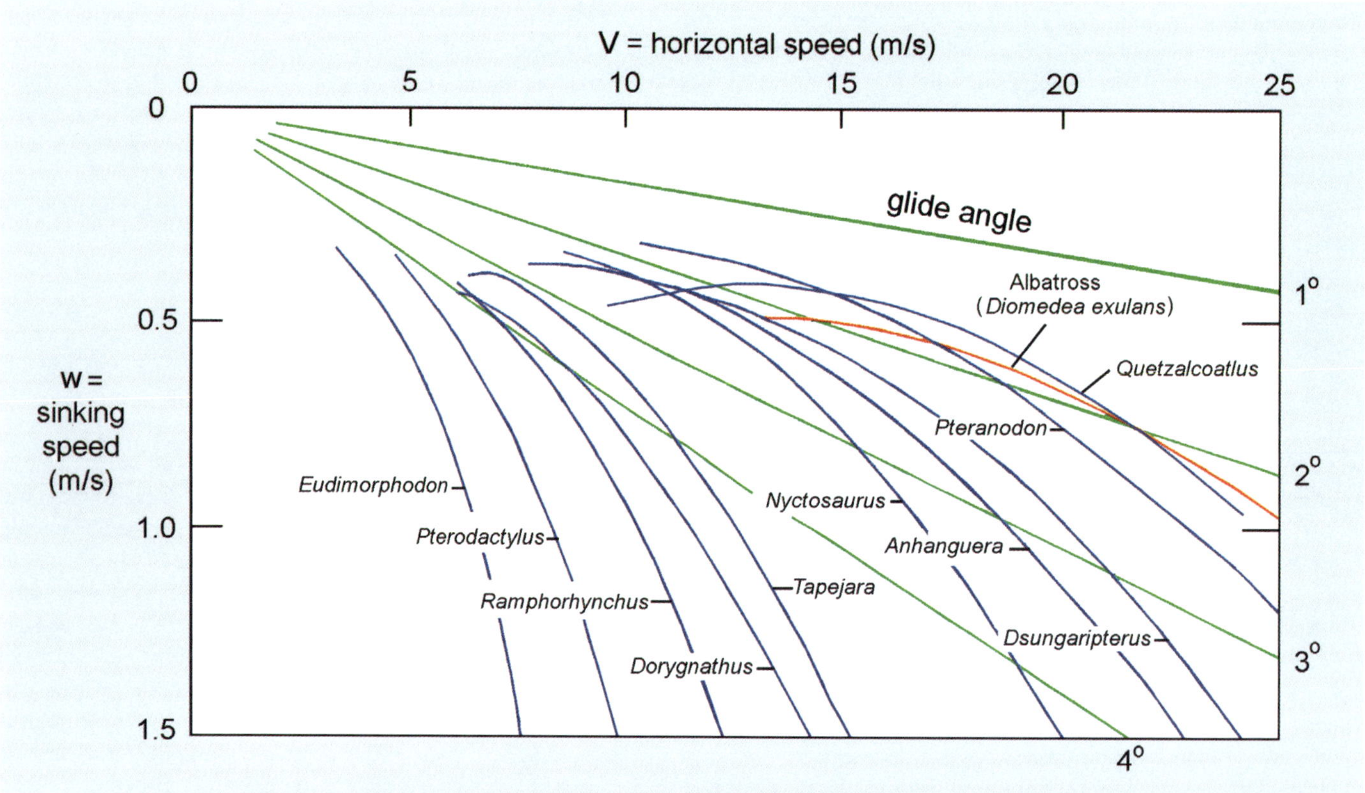

Figure 26. Glide polars for pterosaurs (blue curve) and albatross (red curve) where sinking speed (w) is plotted against horizontal speed (V). The sharp downturn on the left side of each glide polar indicates a stall. The high point of the polar is the minimum sinking speed (i.e., the speed at which the flyer can glide for the longest duration from a given altitude). Flying animals normally glide at speeds faster than the minimum sinking speed so as to avoid stalling. The third point on the right of each polar represents the airspeed at the best gliding ratio. The speed for the best gliding angle (maximum lift/drag ratio) can be found by drawing a tangent between the glide polar and the origin. A glider sinks rapidly in still air if it is traveling slowly or fast, but sinks less rapidly at intermediate speeds. The glide polar curves show that there is a regular progression of gliding performance in pterosaurs as size increases; the larger the size of a pterosaur, the better the gliding performance. The gliding ability of *Quetzalcoatlus* can be compared with that of modern albatross. See Tables 3 and 4 for aerodynamic data.

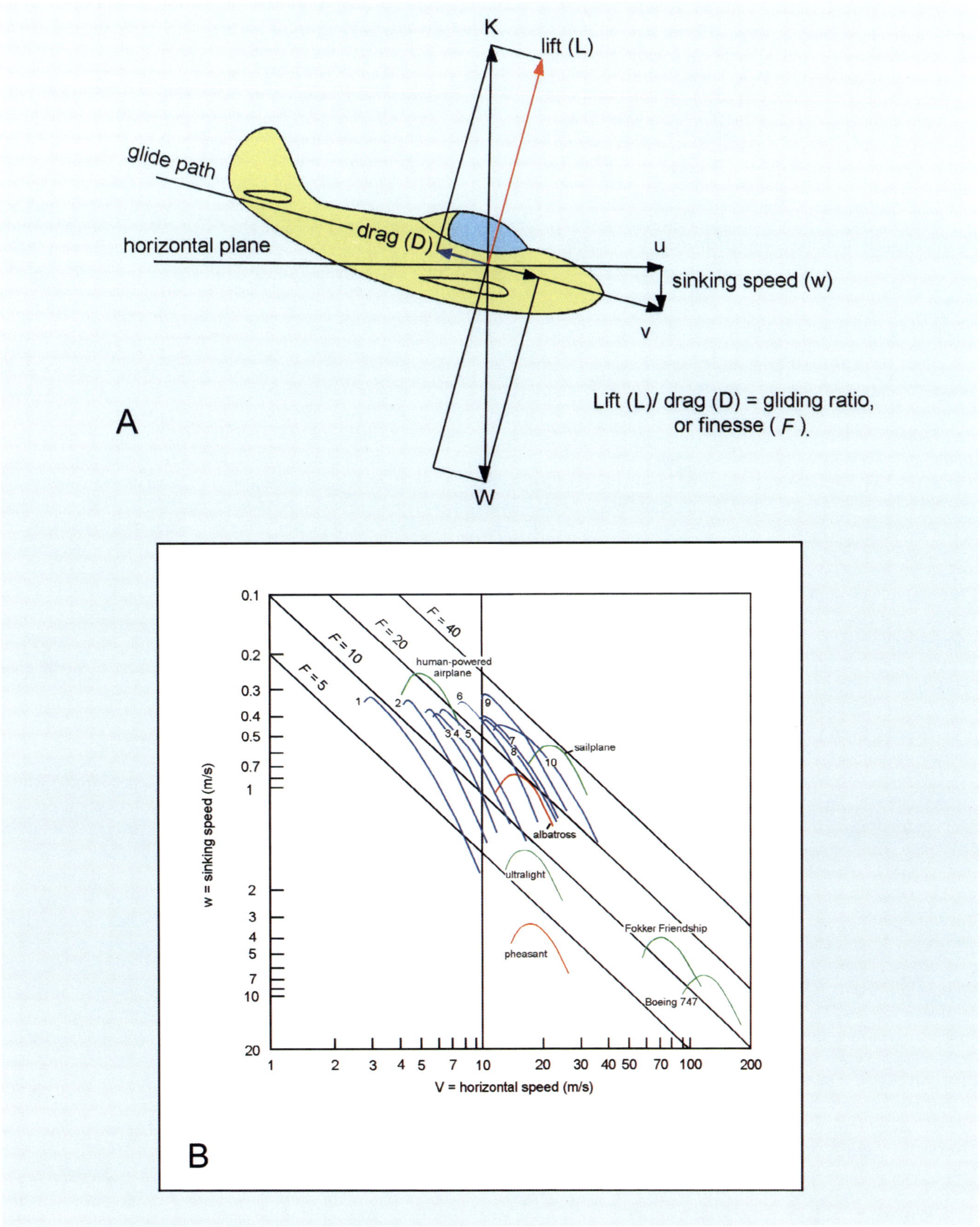

Figure 27. Gliding flight of pterosaurs and other flyers. A: In a glider, the opposing forces are compensated so that $D/W = w/V$, where D is the drag, W is the weight of the glider, w is the sinking speed and V is the airspeed. The ratio L/D between drag and lift determines the gliding ratio or finesse (F) (modified from Tennekes, 1996). B: Gliding performance of pterosaurs and other flyers, gliding at various speeds. Sinking speed (w) of ten pterosaurs and other flyers (albatross, pheasant, ultralight, sailplane, Fokker Friendship, and Boeing 747) is plotted against the airspeed (V) in log-log coordinates. It appears that most pterosaurs were excellent soarers with sinking speed close to 0.3 m/sec, comparable to albatross, human-powered airplane, and sailplane (modified and expanded from Tennekes [1996] by incorporating pterosaur data as shown by blue curves). See Tables 3 and 4 for aerodynamic data.

sinking speeds close to 0.3 m/sec. They were capable of spending a long time over the seas effortlessly by manipulating the wind and the air currents off the waves. Like modern seabirds, they probably used various techniques such as slope soaring and thermal soaring (Brower, 1983), and possibly dynamic soaring with their long and narrow wings.

The finesse (F) of the wandering albatross is ~25, better than that of human-powered airplane. The finesse of most medium-to-large-sized pterosaurs ranges from 20 to 40. There was a certain angle of attack of the wings, which generated the greatest lift-to-drag ratio. Moreover, their rate of descent is excellent, 0.3 m/sec, making them efficient gliders. Advanced sailplanes approach F-values close to 40. In contrast, pheasants and ultralights have poor F-values. Similarly, aircraft like the Fokker Friendship and Boeing 747 with moderately high finesse (between 10 and 20) are quite good power-off gliders when engines fail. The finesse curve indicates that the small pterosaurs have lower gliding airspeeds with their gliding angle close to 4°, but large pterosaurs had excellent gliding performance with high finesse and low gliding angle close to 1°. They could glide long distances with very little descent. Their gliding performance is comparable to that of an albatross, human-powered plane, and sailplane (Fig. 27B). Pterosaurs were probably able to lock the wing in a horizontal position against the lift forces during gliding by pulling the wings down by the caudal part of the pectoralis muscle.

Soaring

Soaring flight is gliding in circles where the height is maintained or gained by rising air. Soaring birds take advantage of updrafts from wind currents. Soaring flight is considerably cheaper than flapping flight. Many large birds use soaring when searching for food and during migration. It is likely that large pterodactyloids took advantage of soaring flight wherever they could, because of the energy costs of flapping. One of the commonest sources of rising air exploited by soarers is thermals or rising columns of warm air (Alexander, 1992). Thermals are produced when the sun heats the ground unevenly. For example, a large, exposed rock surface will warm up more rapidly than the surrounding vegetation (Fig. 28A). As a result, air above the rock surface is warmed differentially and begins to rise as a column. Soaring birds use such landmarks to find thermals. Another signpost is a canopy of cumulus cloud over one or more thermals, which are visible from long distances. Once a bird locates a thermal, other birds converge on it. The bird circles and climbs vertically in the rising thermal; then it glides straight to the next thermal, and so on. Soarers use thermals of sufficient strength to stay airborne without flapping their wings. Thermals have a finite diameter, so soaring birds continually circle within the updraft while they forage, migrate, or just for fun. Turning in sharp circles requires an increase in lift, but this in turn increases aerodynamic drag and increases sinking speed. This is why birds climb in a thermal to some substantial height; they glide off toward another thermal while losing height, and then climb again in a new thermal. Thermals are active during the day, especially between the late mornings to early evening, as they are powered by sun, but die out later in the evening.

Thermal soaring would be crucial for pterodactyloids that ventured inland, such as *Dsungaripterus* and *Quetzalcoatlus*, in search of food (Fig. 28A). It is likely that these pterodactyloids employed cross-country soaring to gain sufficient height before they started gliding. With their low sinking speed (0.4 m/sec), pterosaurs were adapted for flying in small amounts of lift generated by thermals. Thermals produced by small temperature differences (<2 °C) would be sufficient to provide the necessary amounts of lift for pterosaurs (Brower, 1983).

Many soaring seabirds, such as albatrosses, frigatebirds, and gulls, have narrow, tapered wings, high aspect ratios, low camber, and very low induced power requirements similar to those of pterosaurs. More than 90% of fossil pterosaurs have been found in marine shelf beds, indicating that they lived and died in coastal areas and occupied ecological niches similar to those of modern seabirds. Convection currents are common over the oceans. When a wind is blowing, it overtakes the waves and forms an updraft. Also, thermals often form over the oceans at the interface of cold air over warm water. During the summer in the tropics, the trade winds bring in cool air over warm water, whereas during winter at high latitudes cool air from the polar regions flows over relatively warm sea. Such thermals remain active at night because the sea remains warm while land cools (Burton, 1990). This is why large seabirds are restricted to trade wind zones where soaring conditions are optimal throughout the year. Brower (1983) concluded that *Nyctosaurus* and *Pteranodon* utilized these convection currents to soar over ocean and traveled 500 km from the closest shoreline (Fig. 28B). The coastlines may be relatively flat or bounded by slopes of low hills, dunes, and small ridges, but the differences between land and water will create thermals. An efficient method of soaring employed by seabirds is slope soaring when offshore horizontal winds are deflected upwards against cliffs (Fig. 28C). The winds rebound again behind the cliff, creating repeated opportunities for slope soaring. Many seabirds utilize this updraft to soar over a cliff. Upon reaching the end of the slope, they turn around and regain the lift.

Slope soaring is not confined to cliffs but also works over open ocean, on the windy side of the large waves. It generates updrafts that many seabirds, such as storm petrels, fulmars, shearwaters, and gannets, exploit (Pennycuick, 1983). They fly within a shallow layer of moving air over undulating sea waves. Sea waves are generated and maintained largely by wind actions. The wind speeds above the surface may vary from 15 to 40 km/h. It is likely that pterosaurs used the same technique of slope soaring (Fig. 28D) as in modern seabirds (Brower, 1983). Magnificent frigate birds have the longest wings relative to weight of any bird. Their wing design is similar to that of pterosaurs with negative dihedrals, where the wingtips are angled down as in pterosaurs. This makes the animal unstable but gives them much greater maneuverability for chasing other birds for food piracy. Frigate birds endlessly soar high over water, pirate fish from other birds or pluck them from the water, and yet never land on its surface.

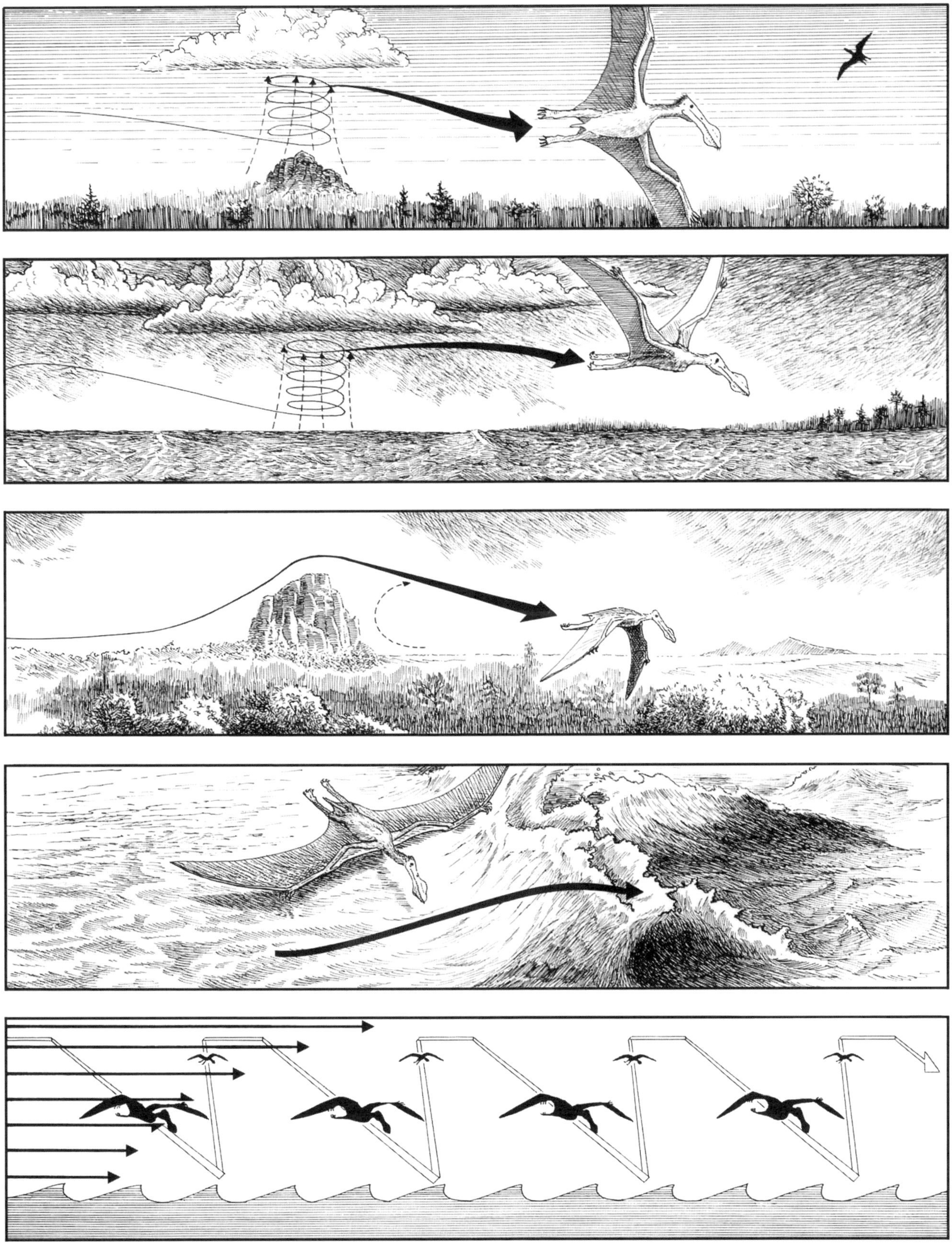

A

B

C

D

E

Figure 28. Various soaring techniques of pterosaurs using lift produced by (A) thermals, (B) vertical convection currents over water, (C) slopes, (D) declivity currents, and (E) dynamic soaring.

Awkward on land, they nest in bushes and trees on oceanic shores, using the slight elevation to gain the needed speed to get airborne. This bird's behavior may be indicative of the lifestyle of medium-sized pterosaurs. A gliding bird may accelerate its speed by flexing and sweeping back the wing like an arrowhead. It is likely that pterosaurs assumed this arrowhead configuration of the wing when needed to accelerate so that the tips aimed back and the knuckle joint formed the point of the arrow (Fig. 17B).

The long soaring flight of the albatross has intrigued naturalists and explorers for more than two centuries. The albatross employs a novel method of soaring, called dynamic soaring, in oceanic regions with strong continuous, unidirectional winds without flapping wings (Pennycuick, 1982, 1983). It has narrow wings like pterosaurs, spanning 2–3 m, with high aspect ratios of 13–15 and a low camber (nearly flat), and relatively weak flight musculatures. Pennycuick discovered a shoulder lock in albatrosses as an energy-saving mechanism to keep the wing in a horizontal position during soaring. Although its humerus is capable of vertical motion during flapping, when pulled forward, it locks forward and cannot be raised. Steady, prevailing winds blow horizontally across great stretches of the sea in layered paths like a deck of cards, and albatrosses follow these wind patterns in their travels. The lowest layer of wind blowing across the ocean moves the slowest, because the friction is greatest near the water surface. The higher the altitude, the faster the wind blows. The albatross exploits this gradient of horizontal wind speed to soar. Over the sea, lift coefficients average 0.9–1.0. The albatross starts out on a steep downwind glide from a height of ~20 m to gain speed, momentum, and kinetic energy. As it loses altitude, it gains groundspeed. At the bottom of the downwind glide, just above the sea level, the albatross turns around and climbs back into the wind. As it climbs higher and higher, the wind layers are moving faster and faster, aiding its lift. Once having regained the altitude, it turns into another downwind glide to avoid stalling. Repeating these upward and downward movements in a zigzag course, an albatross can stay aloft for hours and days without flapping its wings (Kaufmann, 1970).

Is it possible that the largest pterosaurs could have developed the albatross's trick of continuous dynamic soaring in the low-level wing shear, without powered wing flapping? The question was raised by Bramwell and Whitfield (1974) and Brower (1983), but although they did no calculations, they concluded that the flight speed of *Pteranodon* was probably not high enough. They referred to a criterion developed by Walkden (1925), which shows the importance of high airspeed. Walkden gave a simple numerical analysis of dynamic soaring depending on three parameters: gliding ratio, the gliding speed, and the velocity gradient of wind. Since the impressive gliding performance of *Pteranodon* and *Quetzalcoatlus* calculated here is comparable to that of the wandering albatross, it is possible that these large pterodactyloids could duplicate the albatross's ability to soar continuously in a strong wind shear. They probably employed a similar dynamic soaring technique by swooping up and down between faster winds high above the sea and slower winds near its surface, making use of a free energy source (Fig 28E).

HOVERING AND FLAPPING

Powered flight requires much more energy than does gliding. Energy comes from the flight muscles that move the wings up and down to keep the animal aloft. Most functional analyses of powered flight have centered on birds, taking advantage of the sophisticated aerodynamic analyses that engineers use to design aircraft. However, applications of these equations to study the flight performance of birds are not always straightforward. The properties of bird wings are fundamentally different from those of aircraft. For example, bird wings flap, whereas airplane wings are fixed. Bird wings are porous, whereas as airplane wings are solid. Bird wings yield to air pressure, whereas airplane wings resist it (Kardong, 2002). In spite of these differences, simplified assumptions from aerodynamic theory have led to important insights into flight mechanisms.

Tucker (1968) pioneered in studying the flight power of birds in a wind tunnel experiment where he measured the metabolic rate in flight. Flight power is the rate at which the bird consumes metabolic energy and generates mechanical work as it flies. Tucker trained a budgerigar to fly in a specially built wind tunnel and calculated how much energy the bird used at various airspeeds. He calculated the aerodynamic power (P) required to fly in the wind tunnel at various speeds (V) and plotted P against V. He found metabolic energy costs ranging from about 50 watts/kg to 120 watts/kg on various birds and bats. These U-shaped curves are standard power curves for estimating flight performance. He found that in horizontal flight, the most economical condition is at the middle of the speed range. It takes more power when it flies slowly or fast.

Power Required and Power Available

The key to understanding why small pterosaurs hovered, intermediate ones flapped, and large ones soared relies on the margin between the power required for flight and the power available for that purpose. Steady level, flapping flight requires the application of mechanical power to generate lift and to overcome the resistance of aerodynamic drag. The drag force on a bird consists of two components: the parasitic drag due to friction and the induced drag due to lift. For any flying animal, fixed-wing aircraft, or helicopter, the curve of required power against airspeed is U-shaped, being higher at both ends than at the middle. This means that it is harder to fly at low speeds or to hover steadily than it is to cruise at some intermediate speed, and harder again at high speed. This U-shaped curve has a profound effect on all aspects of flight performance. Its shape is typical of helicopters and fixed-wing aircraft, and it has also been found in birds that have been trained to fly in wind tunnels (Tucker, 1968; Dial et al., 1997).

We plotted the power curves of ten species of pterosaurs and of an albatross for comparison (Fig. 29). The basal metabolic rate in birds is roughly 20 watts per kg of body weight. The maximum mechanical power that can be delivered continuously by the wing

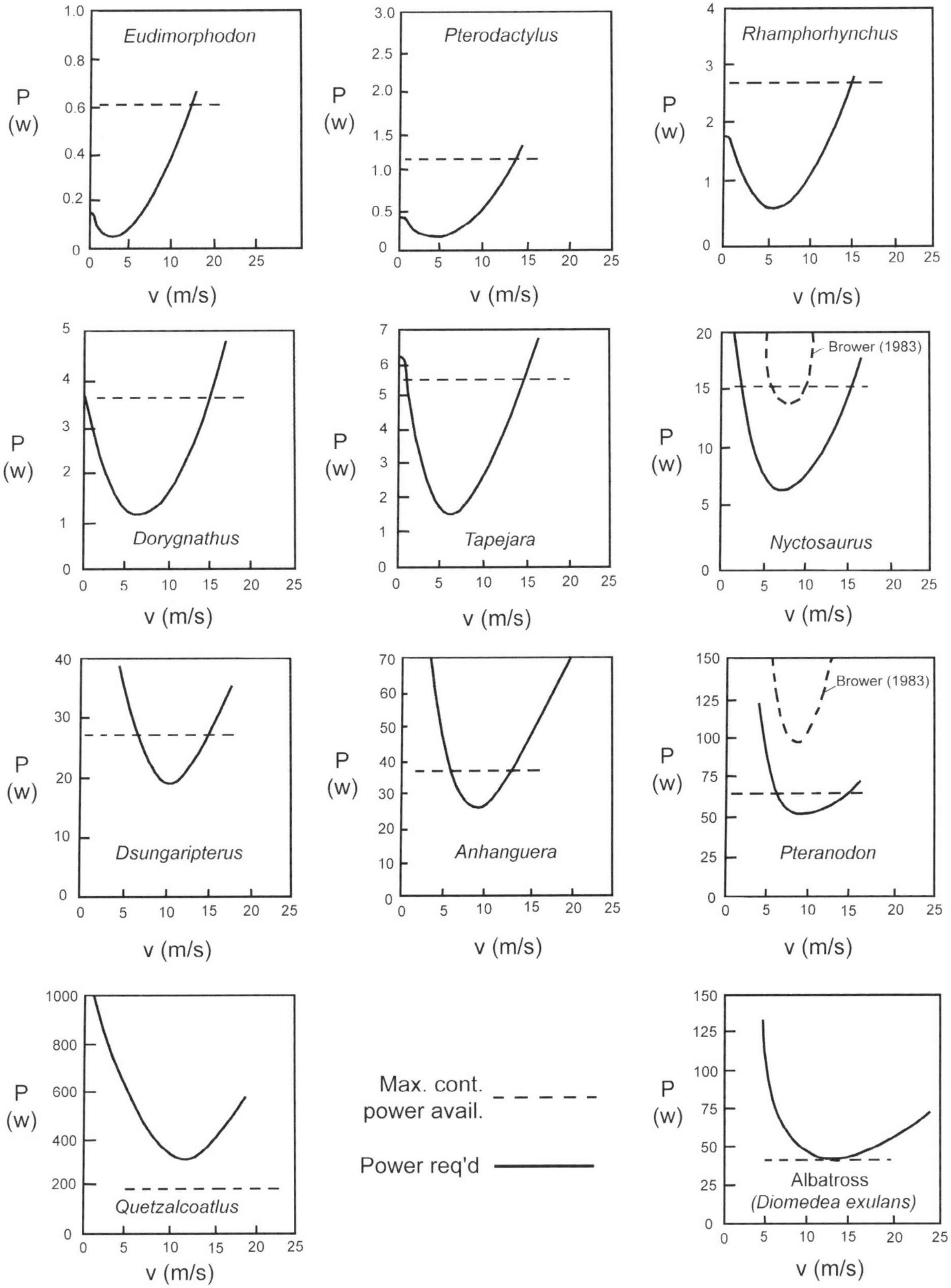

Figure 29. Power profiles for ten genera of pterosaurs and albatross for hovering and flapping flight. The U-shaped curve gives the metabolic power necessary to fly. The left-hand arm of the U-shaped curve is mainly the penalty that must be paid for the production of lift. The trough of the curve represents the minimum power speed where the pterosaur could fly for the least energy expenditure. A tangent line from the trough shows best cruising speed. The computer program ANFLTPWR is used for computation of the power required for steady level flight at specified speeds. It also gives the maximum continuous power available (P_{avail}) according to Equation 20. The dashed horizontal line represents the maximum continuous power available. Three distinct styles of flight emerged from this analysis: hovering flight for small pterosaurs, flapping flight for intermediate pterosaurs, and formation flight for large pterosaurs (see Figs. 21, 25A for comparisons). For *Quetzalcoatlus*, continuous flapping flight was not possible.

muscles of a flying animal is assumed in this study to be 20% of the animal's maximum continuous metabolic rate. The 20% conversion efficiency from the metabolic "heat engine" to muscle power output is within the range quoted in other studies of animal locomotion (Tucker, 1968; Dial et al., 1997). It is also assumed here that the maximum metabolic rate depends only on animal mass (and is independent, for example, of flight speed).

The left-hand arm of the U-shaped power curve is mainly the penalty that must be paid for the production of lift, which supports the animal's weight in level flight. This "induced" power is highest during hovering and decreases as speed increases. The lowest point of the curve represents the minimum power speed where the pterosaur could fly for the least energy expenditure. The best cruising speed is the speed that requires the least energy for a given distance traveled and is higher than the speed for the minimum power. It can be estimated by drawing a tangent from the origin to the power curve. The rising right-hand arm of the U is mostly the power required to overcome so-called parasite drag. In each curve, a horizontal line indicates the mechanical power available for flapping. It is proportional to an animal's continuous metabolic rate. In Figure 29, there are small graphs showing curves of power required (labeled P_{reqd}) in level flight at low altitude for ten species of pterosaurs, from the smallest to the largest. All have the typical U-shaped curves, which are higher at the ends than in the middle. Also shown in each graph is a dashed horizontal line, which is the mechanical power for flapping propulsion. Much greater power than the values shown (more than twice as high, see Brower, 1983) can be developed for short periods, but the values shown here are for sustained flight only. It will be noticed that the maximum steady level speed, where the right-hand branch of the U-curve intersects the available power, does not change radically as size increases in pterosaurs, remaining usually in the range from ~10 to 15 m/s. Three distinct styles of pterosaur flight emerged from this analysis.

Hovering Flight

Many small birds, such as hummingbirds, beat their wings up to 100 times per second to hover while sipping nectar. Hovering is the most expensive form of flying, where small birds and bats hover for a few seconds to pick up food. Flying to stay in one place, while supporting body weight, is a difficult feat. In hovering flight, the main wing movement is centered at the shoulder joint, while the entire wing remains stiff. The body is tilted up at a steep angle so that the wings sweep back and forth in a nearly horizontal plane similar to insect flight (Alexander, 1992). It appears from the power curves that for smaller pterosaurs, such as *Eudimorphodon*, *Pterodactylus*, *Rhamphorhynchus*, and *Dorygnathus*, the available power (P) exceeds the required power (P_{reqd}) at zero speed and they are evidently capable of extended hovering flight, though for *Dorygnathus* the hovering capability was marginal (Fig. 29). The small pterosaurs probably obtained seafood when hovering just over the surface, scooping fish and crustaceans from the water.

Steady Level Flight

In powered flight, flapping wings provide both lift and thrust. In a flapping wing, however, each section performs a separate task. The distal segment of the wings is mainly responsible for flapping strokes. It involves a downward and forward power stroke with the wing fully extended to provide thrust, followed by an upward and backward stroke with the wing partially retracted (Rayner, 1988). With a complete wing beat cycle, the wing creates a lazy figure eight. The tips of the wings not only flap up and down but also twist forward on the downstroke to propel the bird forward. The upstroke is largely a recovery stroke. It serves to position the wing for the subsequent downstroke. The medial segment of the wing provides lift to overcome gravity, while the distal segment of the wings generates forward thrust to overcome drag. Lift is achieved by the airfoil action of the wing, and thrust by flapping the wings up and down in a complex manner.

The most common style of flight adopted by small to medium-sized pterosaurs was probably steady level flight. The slightly cambered and thin wings of pterosaurs with high aspect ratios and tapered tips indicate their adaptation for fast flight. They had rapid wing beats and thin, streamlined bodies to reduce parasitic drag, which is high at fast speeds. As pterosaurs got larger, their wing loadings increased, and their wing-beat frequencies decreased. It appears from the power curves that all pterosaurs except *Quatzalcoatlus* were capable of steady level flight to surprisingly similar maximum speeds around 15 m/s at sea level, though *Pteranodon* is marginal. An exception is the little *Eudimorphodon*, which reaches about 10 m/s. *Quetzalcoatlus* needed about 600 W to fly level, which is about 3 times its maximum aerobic power (Fig. 29).

Formation Flight

It is possible that larger pterosaurs such as *Dsungaripterus*, *Anhanguera*, *Pteranodon,* and *Quetzalcoatlus* used another strategy to reduce the energy cost of flight during long-distance migration, namely formation flight. Many modern birds, such as geese, ducks, pelicans, and others, fly in formation. In this mode, each animal follows the leader with its wingtip slightly overlapping the opposite tip of the animal ahead, and trailing behind, so the formation is either V-shaped or in single echelon line. All except the leading animal receive the same benefit from the upwash outboard of the trailing vortex from the animal immediately ahead.

The accurate estimation of the drag and power reduction obtainable by flying in formation is difficult, owing to the complexity of the cyclic nature of the vortex field shed behind flapping wings. Here we have made what can only be considered an order-of-magnitude estimate, based on the assumption that the prominent vortices are those shed in a steamwise direction behind the wingtips, like those visible as vapor trails behind the jet aircraft at high altitudes. There is an induced "upwash" immediately outboard of the cores of these vortices, with a strength related to the animal span loading Mg/b, the flight speed V, and of course to the location of each animal in the formation. The upwash intensity will vary considerably across the wingspan of the follower, which must force it to make some adjustment in order to prevent upset, but its effect is similar to the uplift obtained when soaring in an upward thermal flow. The upward velocity component is assumed to tilt the lift vector forward relative to the horizontal flight path, providing an effective reduction in aerodynamic drag and required power. Without describing the mathematical details, we have estimated the power reduction to be approximately:

$$\Delta P = \frac{-(\mathrm{Mg})^2}{\pi \rho e b^2 V} \tag{22}$$

In the SI metric system the units of power are watts (symbol W).

We have shown the power curves for formation flight for the four heaviest pterosaurs, namely *Dsungaripterus*, *Anhanguera*, *Pteranodon,* and *Quetzalcoatlus*, by the left-handed set of numbers 7 to 10 (Fig. 30). Note that the power reduction is greatest at lowest speeds (marked by diamonds), which are reduced

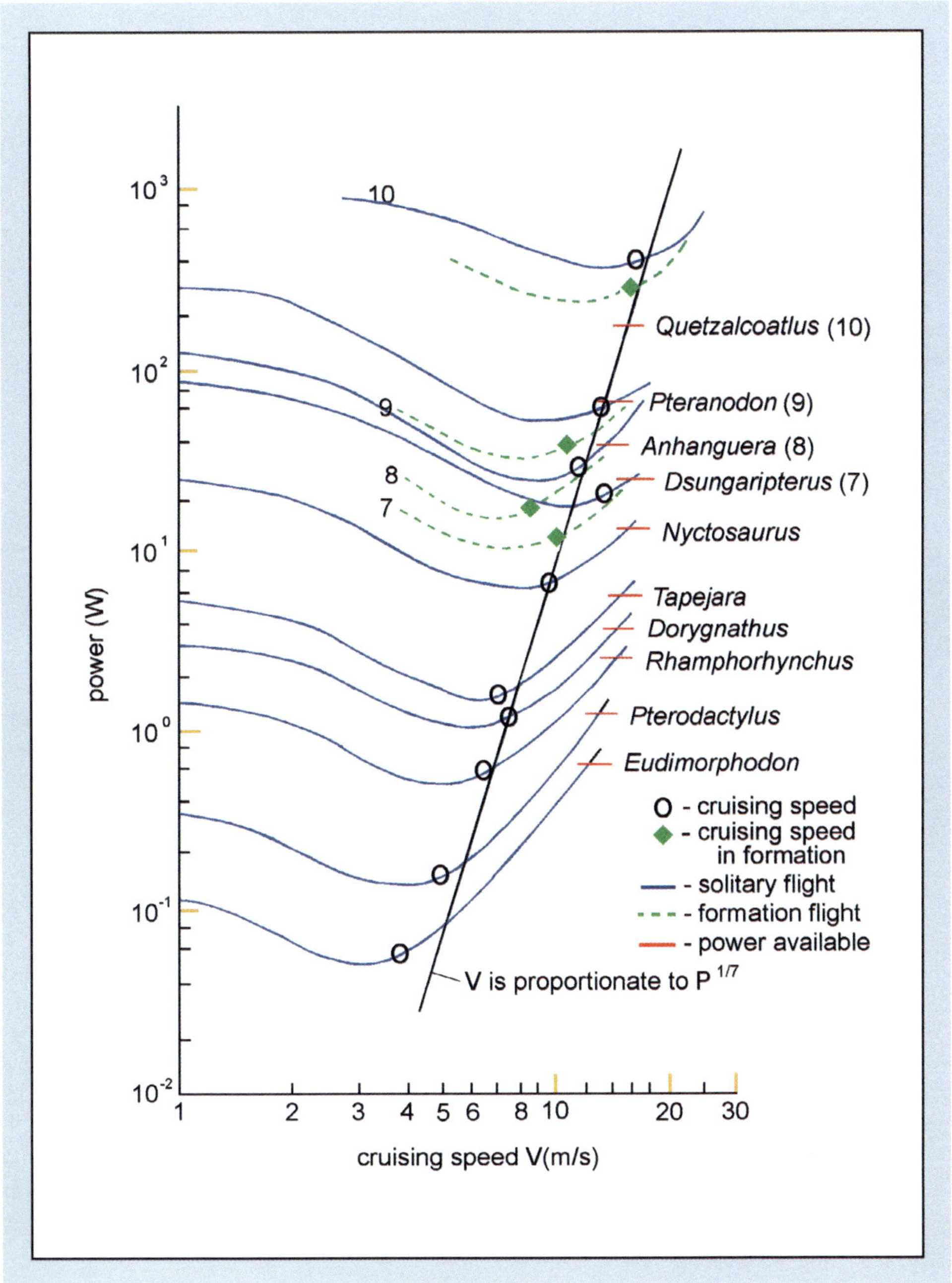

Figure 30. Power profiles for ten pterosaur genera; power (W) when plotted against cruising speed (V) on logarithmic coordinates, tends to show an interesting pattern. The power curves are neatly arranged showing a nearly linear (power-law) relation between power at maximum L/D ratio (the best cruising condition) and cruising speed, where $V \propto P^{1/7}$.

somewhat below that in solitary flight. The chart shows that, in some cases, the best cruising speed in formation is about the same as the speed for the minimum power for a solitary flyer (e.g., the leader of the flock). Since the leader may be hard-pressed for power, it may want to stay near the minimum power speed, and then the followers will automatically be near the best cruising speed. Since formation flight power reduction is greatest at low flight speeds, the best cruising speed is slower in formation than in solitary flight. There is a rotational change of the leader of the group, who experiences more drag and needs more energy. Evidently, formation flight may not permit *Quetzalcoatlus* to fly continuously. It requires so much anaerobic power for solitary flight that *Quetzalcoatlus* could never lead a migrating flock. The power profile shows that *Quetzalcoatlus* was an obligate soarer but poor flapper.

When we plot the power curves on log-log scales against speed, an interesting picture emerges (Fig. 30). The power curves are neatly arranged, showing a nearly linear (power-law) relation between power at maximum L/D ratio (the best cruising condition) and cruising speed where

$$V \infty P^{\frac{1}{7}} \quad (23)$$

Here we try to explain why this allometric relation exists between speed and power. It can be shown that that, for geometrically similar gliding animals, the lift and drag coefficients at maximum L/D and the maximum L/D ratio itself are nearly constant, irrespective of size. Then the power required is proportional to $V^3 \times S$, where S is wing area. Now it can also be shown that at fixed lift coefficient (C_L), the speed V is proportional to the square root of the wing loading M/S, but since M varies as $S^{3/2}$, M/S varies as $S^{1/2}$, and V must vary as $S^{1/4}$, and thus S as V^4. So finally the power at (L/D) maximum is proportional to $V^4 \times V^3 = V^7$, and V ought to be proportional to $P^{1/7}$. The cruising speed for *Eudimorphodon* is lower than a 1/7 power relation would suggest, which is due to somewhat higher viscous drag at the small scale of this animal. In the computation, the effective skin friction drag is a function of Reynolds number.

TAKEOFF AND LANDING

Takeoff

As with birds, the most crucial maneuvers in pterosaur flight seems to be takeoff and landing. Takeoff requires more energy than does level flight because the bird must accelerate and climb and because takeoff speed is near the left-hand end of the U-shaped power curve. Whenever birds with mass greater than about 0.1 kg take off from the ground, they need speed to become airborne. The bird must beat its wings more vigorously to obtain the lift it needs. Takeoff from the ground is hard work that often requires a short burst of anaerobic power. For this reason, most birds prefer to take off from an elevated perch to secure the needed lift. It is likely that pterosaurs might have adopted a similar strategy. A cliff-dwelling pterosaur, or one on a tree, merely had to jump into the air and spread its wings; the force of gravity would supply the necessary speed to achieve lift for flight.

When birds take off from a perch, they do not seem to use excess power; they lose height at first and then swoop up with a large amplitude undulation. This occurs whenever any winged object (aircraft, model glider, or flying animal) finds itself in a non-equilibrium situation, such as being launched without sufficient wing lift to balance its weight (Templin, 2000; Chatterjee and Templin, 2003). The result is an initial loss of height and increasing speed. Lift increases as the square of speed (if the control surfaces are not moved) and the subsequent motion is an undulation, known as a phugoid oscillation, in which potential and kinetic energy are periodically exchanged. In gliding flight, the motion is eventually damped to a steady glide, and in fact, the rate of damping is inversely proportional to the lift/drag (L/D) ratio. Pterosaurs with high L/D configuration had low phugoid damping, but because the period of motion is proportional to speed, control is not difficult.

In Fig. 31, we have shown glide paths of ten pterosaur species computed from ANFLTSIM with a launch speed of 3 m/s (horizontal) from an arbitrary height of 20 m. Moderate pitch damping was used in order to suppress large phugoid amplitudes. All pterosaurs could take off easily from the cliff with no wind,

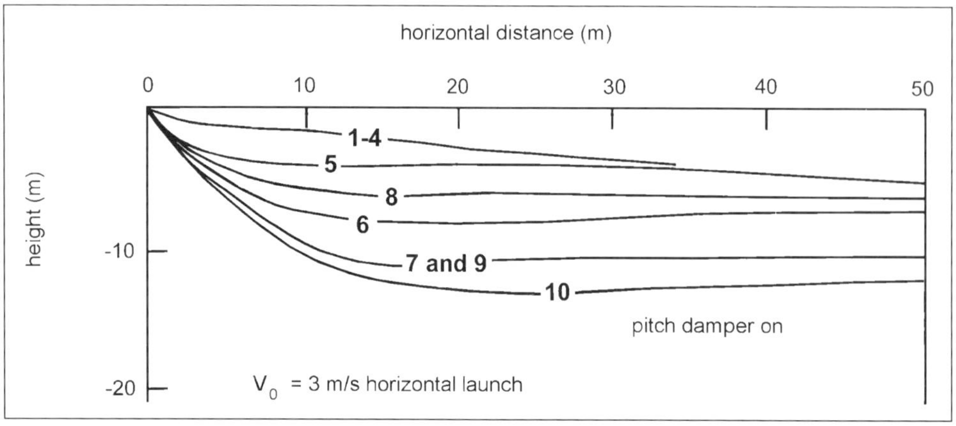

Figure 31. Flight paths of ten pterosaur genera, numbered 1–10 (see Table 1 for names) taking off horizontally from a perch at 3 m/sec, then pulling up a maximum continuous power. Moderate pitch damper (such as a tail) was used to suppress large phugoid oscillation. All pterosaurs could take off from the cliff with no wind, but with greater height loss as mass increases.

but with greater height loss in animals with larger masses. The largest of the pterosaurs is *Quetzalcoatlus*. Following a launch at an initial speed of 3 m/s, with flight attitude trimmed for a high lift coefficient (~1.0), the 70 kg *Quetzalcoatlus* develops insufficient aerodynamic lift to balance its weight, and its path steepens rapidly. As speed builds up, the glide path flattens, and if undampened at a constant lift coefficient of 1.0, it would curve upward to a new low-speed peak and continue to oscillate with a slowly descending altitude, like a roller coaster. The effect of the pitch damper, which increases or decreases the lift coefficient in proportion to longitudinal acceleration or deceleration, reduces the initial height loss, and eventually stabilizes the path to an equilibrium glide at 13.4 m/s and about –2° slope.

However, it is likely that pterosaurs would routinely land on water or ground during feeding. How did they take off from the ground in calm and windy conditions? Many small birds can take off from the ground in still air by flapping their wings to generate enough lift, which enables them to climb steeply. In these birds, the supracoracoideus muscle, the main wing elevator, is powerful. Other birds need to run to gain airspeed. Seabirds such as albatrosses, with their long, high-aspect ratio wings, are awkward during takeoff. They must run for some distance before they can take off, especially when there is no headwind. For them, standing takeoff is more difficult. They cannot flap vigorously for lift generation without damaging their wings if they are standing still. They can take off only with a long run and slow climb. The hindlimbs play a major role to gain the necessary speed until the wings take over to flap vigorously. These birds need long and wide runways during takeoff. A strong wind helps them to take off in coastal regions, where they can climb into the wind and catch updrafts from waves to gain extra lift.

Could pterosaurs take off from the ground, especially when they ventured inland? They had relatively low wing loadings compared to that of seabirds that may be advantageous during takeoff and landing (Fig. 22). To take off from the ground, it would have been necessary for them to shift their posture from the quadrupedal to a bipedal mode to gain the necessary speed. It is known that similarly sized bipeds and quadrupeds use nearly the same amount of metabolic energy to run, despite a dramatic shift in posture and running mechanics (Roberts et al., 1998). Thulborn (1990) suggested a simple allometric equation where the running speed (V_{max} in km/h) can be calculated from the hip length (h in cm); the latter is about 95% of the combined length of femur and tibia:

$$V_{max} = (6.12h)^{0.5} \tag{24}$$

Equation 18 could be written (h in cm) as

$$V_{max}\left(\frac{m}{s}\right) = (0.47h)^{0.5} \tag{25}$$

Using this equation, we calculated the running speeds of ten species of pterosaurs (Table 6).

Like any flying object, the heavier a pterosaur was, the more trouble it had during takeoff and landing. Smaller pterosaurs could take off easily after a short run from the ground with or without a headwind. Large pterosaurs, especially *Pteranodon* and *Quetzalcoatlus*, must have used headwinds or sloping surfaces for extra power in addition to a takeoff run. We calculated the ground takeoff energies of *Pteranodon* and *Quetzalcoatlus* using these two variable parameters: with or without headwind (5 m/s) and with or without ground slope (–10°). Using a modified version of ANFLTSIM, we calculated the metabolic power

TABLE 6. ESTIMATION OF RUNNING SPEED OF 10 SPECIES OF PTEROSAURS

Species	Femur (cm)	Tibia (cm)	Femur + Tibia (cm)	Hip height (h) (cm) (95% of F + T)	Maximum Running speed V_{max} (m/s)
1. *Eudimorphodon ranzii*	1.9	2.5	4.4	4.18	1.40
2. *Pterodactylus antiquus*	3.47	4.83	8.3	7.9	1.93
3. *Rhamphorhynchus muensteri*	3.3	4.9	8.2	7.79	1.92
4. *Dorygnathus banthensis*	4.5	5.8	10.3	9.79	2.15
5. *Tapejara wellnhoferi*	8.82	11.77	20.59	19.57	3.04
6. *Nyctosaurus gracilis*	10.14	16.27	26.41	25.09	3.44
7. *Dsungaripterus weii*	22.1	27	49.1	46.65	4.69
8. *Anhanguera piscator*	23.2	28.2	51.4	48.83	4.80
9. *Pteranodon longiceps*	26.9	42.83	69.73	66.24	5.59
10. *Quetzalcoatlus northropi*	89.9	134	223.9	212.7	10.02

(*P*) needed to take off, liftoff distance, and liftoff speed for the ten pterosaur species, where asterisks (*) in the power column indicate anaerobic power (Table 5).

It appears from Table 5 that small and medium-sized pterosaurs had sufficient available power to take off easily from the horizontal ground with a short run, especially with some wind. They all could take off in zero distances with a headwind of 5 m/s.

Because of their long wings that were rooted to their hindlimbs, pterosaurs could not flap during takeoff. Long wings and short legs were an impediment to rapid flapping as they would provide small amplitudes. Moreover, in this upright posture, the glenoid has rotated almost 60° from the flight position. In this position, when the wing is extended, the palm is facing forward in pronated position. As a result, flapping would not produce any lift. Moreover, for large pterodatyloids such as *Quetzalcoatlus,* long and wide runways would be essential for safe takeoffs so that wings were also not damaged by vegetation. Such wide spaces could be available near coasts, but not in the wooded Big Bend area, where they frequently visited. It is likely that pterosaurs would delay opening the wings during takeoffs until the legs were retracted, a strategy used by many insects. Most likely they would fold their wings during bipedal running to reduce drag and gain momentum (Fig. 32). As soon as they attained flying speed, they would leap into the wind from a crouched position and spread their wings and legs using the air movement to obtain lift. During leaping, the knee and the ankle would be flexed in a Z-configuration and the large head would be tilted backward so that maximum extension could be gained at takeoff. Flexing the ankle joint would stretch the tendon of gastrocnemius muscle, which ran behind the long tibia and attached to the ankle. The elastic energy stored in the tendon would help to catapult the body into the air during takeoff. A great deal of timing and

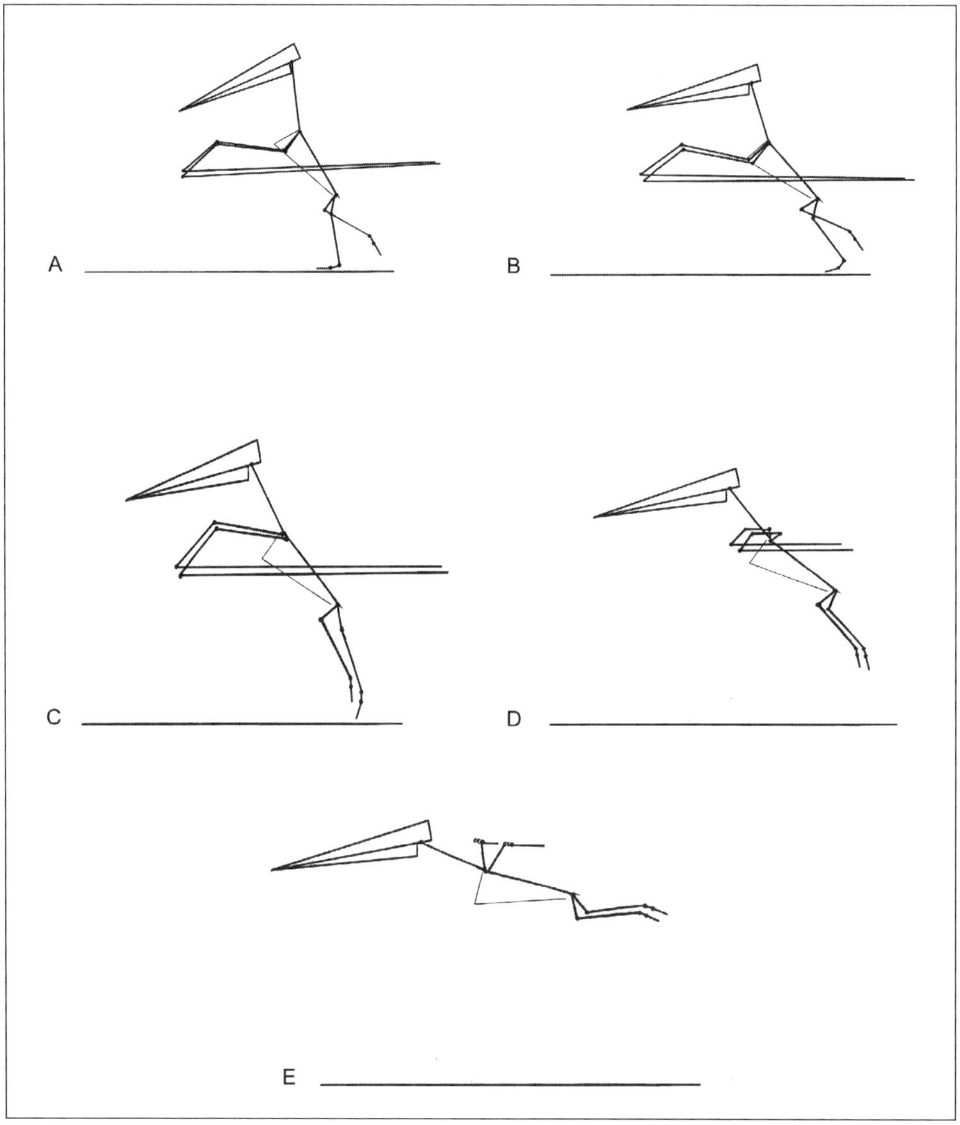

Figure 32. Stick diagrams showing the takeoff sequences of *Anhanguera* from bipedal running position (A) to final airborne stage (E).

coordination would be necessary between the retraction of the legs and the opening of the wings during takeoff. It is possible that once the animal has reached its running speed, sufficient aerodynamic thrust could have been produced by low-amplitude flapping at least with maximum aerobic power, without producing much lift until takeoff speed was reached, followed by a modest jump and increased lift to clear the ground. Leaping would clear the wing from the ground for vigorous flapping in midair. As the ballistic path of leaping reached a very steep angle for free fall, the pterosaurs would extend their wings to flap and gain lift. The spine had now assumed a near-horizontal attitude. The high-velocity rotation of the humerus around its longitudinal axis imparted by the supracoracoideus muscle would be responsible for positioning the wing in the upstroke position (Poore et al., 1997). The takeoff sequences of *Anhanguera* were initiated by running bipedally with wings folded, then crouching and leaping into the air at a 45° angle (Fig. 32).

The zero-wind takeoff distances listed in Tables 5 and 7 have been computed on the assumption that wings may be folded during the ground run until maximum running speeds are attained, and that only limited flapping amplitudes would be needed thereafter until speed is sufficient for liftoff. During this final acceleration phase (liftoff speeds are not much greater than running speeds in Table 5), the pterosaur uses its legs to provide partial support only. The alternative strategy of leaping into the air with suddenly extended wings at the end of the ground run, thus avoiding wing flapping near the ground, is discussed later.

Even with the help of a headwind, taking off would be most strenuous for large pterodactyloids such as *Pteranodon* (16.6 kg) and *Quetzalcoatlus* (70 kg). Avian flight muscle contains a mixture of fast glycolytic and fast oxidative glycolytic fibers; the former permits short-burst, whereas the latter allows sustainable flight. Marden (1994) pointed out that supplementing the flight muscles with the more anaerobic (glycolytic) fibers would have provided large pterodactyloids with ample power to take off. Marden cited a modern analog. For example, the North American wild turkey (M = 10 kg), when chased, takes off vertically from a standing start. Such flights are short-burst and are powered by anaerobic metabolism. Large pterodactyloids such as *Pteranodon* and *Quetzalcoatlus* had to depend on brief periods of high anaerobic power generation for takeoff.

We define the power Pa as the maximum continuous power available, presumably aerobic power, which builds up no oxygen deficit. Generation of a temporary power level P_{max}, greater than Pa, is limited to a duration T (in seconds) by the production of lactic acid in the muscles (McMahon, 1984). T depends on the ratio P_{max}/Pa. McMahon also gives some data for recovery times after intermittent exertion, which seem to be considerably longer than the aerobic power-on times. Brower (1983), quoting other references, suggests a typical value of $P_{max}/Pa = 40{:}17$ (i.e., 2.35), but gives no corresponding duration. Similarly, Marden (1994) estimated that the maximum anaerobic power outputs for birds are roughly 2–2.5 times the aerobic limit. However, for humans, this ratio can be much higher, perhaps as high as 5 for a few seconds (Avallone and Baumeister, 1987, p. 9–168). Using these data, we fitted the following equation to the data:

$$\frac{P_{max}}{Pa} = 1 + 8.38\left[T(\text{second})\right]^{-0.4} \qquad (26)$$

or its inverse:

$$T(\text{sec}) = \left[\frac{8.38}{\left(\frac{P_{max}}{Pa-1}\right)}\right]^{2.5} \qquad (27)$$

Some numerical examples follow:

P_{max}/Pa	1	1.5	2	2.5	3	3.5	4	4.5	5
Duration T (sec)	inf.	1150	203	73.7	35.9	20.6	13.0	8.9	6.4

We assume that these exhaustion rates hold for all warm-blooded vertebrates, including pterosaurs, and the equation could be scaled up in size merely by using the appropriate value of the

TABLE 7. PTEROSAUR TAKEOFF GROUND RUN TIMES AND DISTANCES

		No wind		5 m/s headwind	
P_{max}/Pa	P_{max} (W)	Liftoff time (sec)	Liftoff distance (m)	Liftoff time (sec)	Liftoff distance (m)
(a) *Pteranodon* (M = 16.6 kg; aerobic Pa = 65 W)					
1.0	65 W	13.7 s	82.9 m	2.3 s	5.25 m
1.5	97.5 W	5.6 s	32.1 m	Immediate liftoff	
2.0	130 W	3.7 s	20.9 m	Immediate liftoff	
(b) *Quetzalcoatlus* (M = 70 kg; aerobic Pa = 170 W)					
2.5	425 W	21.2 s	277 m	0.20 s	2.0 m
3.0	510 W	14.8 s	190 m	0.15 s	1.5 m
3.5	595 W	11.0 s	141 m	Immediate liftoff	

aerobic power (*Pa*) for each animal. Some of these calculations (using ANFLTSIM) for large pterodactyloids such as *Pteranodon* and *Quetzalcoatlus* are summarized in Table 7.

It appears from the above table that *Pteranodon* could take off (*Pa* = 65 watts), perhaps without using anaerobic power, but only with a moderate headwind. However, without a headwind, it was successful at 100 watts with a little help from anaerobic physiology. Brower (1983) suggested that *Pteranodon* needed the additional anaerobic power. On the other hand, *Quetzalcoatlus* needed high levels of anaerobic power (500 watts, about 3 × *Pa* for 10 seconds or 4 × *Pa* for 5 seconds) for reasonable takeoff performance. The above empirical equation for the exhaustion times suggests that 3 × *Pa* might be available for 30 seconds and 4 × *Pa* for 13 seconds. This exceeded the actual power at the assumed running speed (10 m/s) plus 5 m/s headwind, so no further run was needed. However, it is not sufficient merely to lift off. High power must be maintained for as long as possible to gain some altitude in a shallow climb before relaxing. *Quetzalcoatlus,* therefore, needed as much wind as it could get. Interestingly, *Quetzalcoatlus* could take off at a maximum continuous power (170 watts) with no wind from a 10° downward slope and required no power in a 5 m/s headwind.

The ground takeoff strategy of *Quetzalcoatlus* is shown in Fig. 33. From a standing start, *Quetzalcoatlus* accelerates downhill with 500 watts power for 10 seconds. After about 2.8 seconds and a run of 14.5 m, it reaches the takeoff speed of about 8.7 m/s followed by a low-amplitude flapping, and by the time it passes over the end of the 50 m grade, it is at 6.3 m altitude. When the power is cut to zero after 10 seconds, the height is 10.5 m. After the power is cut, there is an immediate downward acceleration due to temporary loss of lift, but with the pitch damper operating to suppress phugoid oscillations, its speed recovers to about 11.6 m/s at 13 seconds after launch and a climb begins with final equilibrium at 11 m/s in a flat forward climb of ~3° slope. This is only one of many possible scenarios for takeoffs from sloping ground. Full power might be left on longer, possibly as long as 30 seconds. In that case, a horizontal distance of 290 m and a height of 19 m are reached after 30 seconds. The subsequent power-off glide is similar to the 10-second power case but with about 20 m greater altitude.

Since, as pointed out earlier, there may be severe limitations on the ability of large pterosaurs to flap their wings during takeoff, a brief study using ANFLTSIM was made of the effect of leaping into the air with powered flapping at the end of the ground run, as sketched in Figure 32. It was assumed in each case that the leap was initiated at 45° upward with initial speed equal to the running speed listed in Table 5. For the three largest pterodactyloids—*Anhanguera*, *Pteranodon*, and *Quetzalcoatlus*—many trials were made, with various settings of power, initial wake angle (equivalent to setting average wing lift coefficient), and pitch damper gain. The pitch damper reduces lift during initial speed reduction in the climb to maximize height and increases lift in the subsequent accelerating descent to rapidly pull up and avoid ground contact, before finally setting into an equilibrium climb. High pitch damping was required to achieve success without exceeding the maximum wing lift limit during the pull-up. The results may be summarized as follows. No takeoff was successful using only maximum continuous (aerobic) power. This was not surprising in the case of *Quetzalcoatlus*, since it is theoretically not capable of steady level flight at this power level (170 W), but the other two have some limited level flight capability with aerobic power (39 W for *Anhanguera* and 65 W for *Pteranodon*). In all cases, success (advance of ground contact) was achieved at anaerobic power levels from about 2.5 to 3 times maximum aerobic power (100 W for *Anhanguera*, 200 W for *Pteranodon*, and 400 W for *Quetzalcoatlus*). The maximum heights reached in the initial arc depended on the running speed, and were about 0.8 m for *Anhanguera*, 1.0 m for *Pteranodon*, and about 2.9 m for *Quetzalcoatlus*. The minimum ground clearance in successful takeoffs was about 0.5 m or less.

Pterosaurs might have used another strategy to take off from the ground. Many birds use ground effect to fly effortlessly close to the surface of water or ground. The effect depends on the ratio of wingspan to height above ground and pretty well disappears if the height is more than about half the wingspan. The presence of the ground limits the amount of "downwash" generated along the wing by the tip vortices. The induced drag is associated

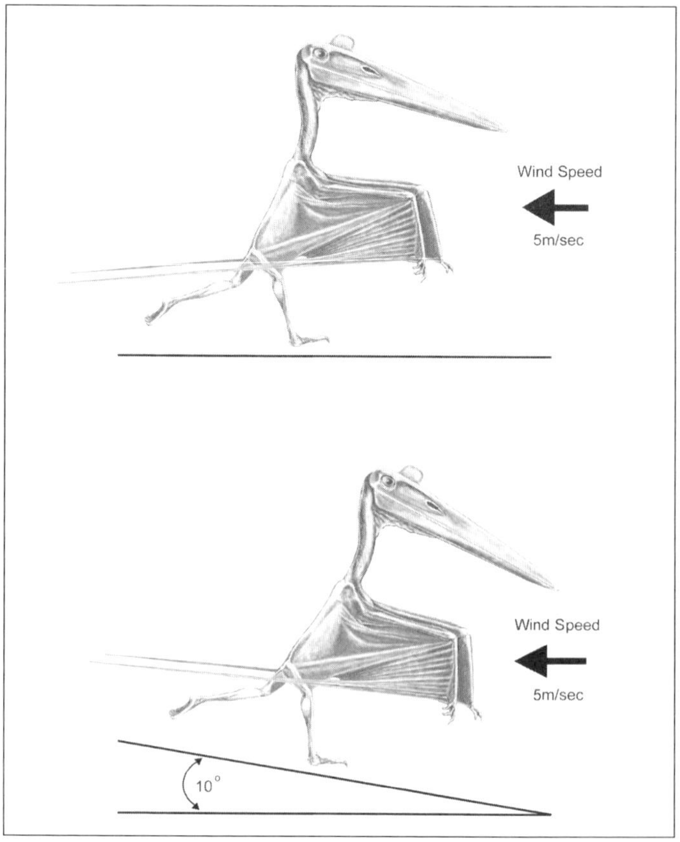

Figure 33. Takeoff strategy of *Quetzalcoatlus* using headwind of 5 m/s and a 10° downslope

with downwash, so for a given amount of lift, the induced drag is reduced, and so is required power (if power is expended). It is likely that large pterodactyloids with long wings might have utilized ground effect. From a running start they would launch into the wind to become airborne before they reached full flying speed. They would keep flying low to use the ground effect to gain enough airspeed until they could climb above the ground effect.

Bramwell and Whitfield (1974) suggested that *Pteranodon* could take off from a wave crest by either flapping its wings or launching itself as a glider if sufficient lift were present. However, they made no calculations to support this notion. We doubt that *Pteranodon* and *Quetzalcoatlus* could take off from water. Wave crests may give some momentary elevation and upward speed for a start from which to launch into the wind. Except for the front face of breaking surf in shallow water, waves do not generally provide much "push." Rather, as the wave passes by, it lifts the object and returns it to nearly the same place. Nevertheless, the wave lift may provide a brief opportunity for vigorous wing flapping at high power (apparently high anaerobic power would be needed in the case of *Quetzalcoatlus*), without slapping the water surface. If the wave motion is more or less in the wind direction, a launch into wind implies that takeoff from a wave crest should be opposite to the wave motion.

Landing

Landing, like takeoff, is an arduous task. All birds land as slowly as possible to keep from injuring themselves. They land into the wind to decrease the ground speed while maintaining enough airspeed for lift. The safe landing speed is about 5 m/s for animals of any size (Templin, 2000). Small pterosaurs could slow down below their minimum cruising speed by tilting up their wings to increase drag and land safely without difficulty. Braking is accomplished not only by raising the angle of attack, but also by tilting the entire body upright, by spreading the uropatagium, and by lowering the legs. By controlling the pteroid bone, the propatagium might act as a flap to reduce speed. Just before touchdown, they would lower their legs forward to absorb the impact of landing. Landing was evidently on two legs because wings would be used as a brake and controlling device. During landing, a headwind was also helpful to slow down, shortening the landing run considerably. Low wing loadings helped pterosaurs to make slow, gentle landings. Slow landing was especially important for them to protect their long wings. Landings on water may have eased the landing shock for small pterosaurs, but water takeoffs for very large pterodactyloids may have been impossible.

Did large pterodactyloids land somewhat awkwardly, like albatrosses? Bramwell and Whitfield (1974) envisaged that *Pteranodon* made a "belly landing," since its legs were too weak to support the impact, and held its wings high to avoid damage. During landing, albatrosses "waggle" their wings by rapidly increasing and decreasing the angle of attack to provide additional braking. If there is no headwind, they land on their feet and use their breasts for additional support (Burton, 1990). With a headwind, they glide slowly and land safely. In the case of pterosaurs during bipedal landing, the slackened uropatagium suspended between the femora could have acted as a shock absorber, as in bats (Vaughan, 1966). In Figure 34, we show two possible landing scenarios for *Quetzalcoatlus*. If the ground is approached in an equilibrium glide at its flattest angle (–1.67° with no wind and –2.28° in a 5 m/s headwind), its airspeed is 15 m/s, well above the stalling speed, which is about 10.9 m/s for C_{Lmax} = 1.2, and still further above the safe landing speed (5 m/s). Can the high gliding speed be rapidly reduced to 5 m/s or less? A deceleration technique used by many birds, flying squirrels, and hang-glider pilots is to level the glide at high speed, then pitch up to high lift with the wings as a kind of lifting horizontal parachute. This is simulated, with and without wind, in Figure 34. The landing speed appears to be excessive with no wind. However, if there was a headwind of 5 m/s, *Quetzalcoatlus* could land safely at a speed of 3 m/s.

SEXUAL DIMORPHISM AND AERODYNAMIC FUNCTION OF THE HEAD

Among flying vertebrates, the most unusual feature of pterosaur anatomy is its immense skull, often dominated by a huge sagittal crest. Did this enormous skull help or hinder the flight performance of the pterosaur? The skull is lightly built, with hollow, paper-thin bones stiffened by struts in critical areas. The function of the crest has been debated for a century. The most popular view is that the crest functioned as a rostral rudder,

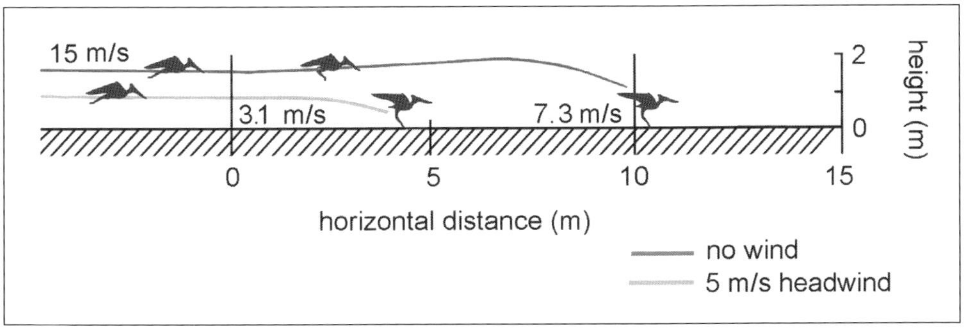

Figure 34. Safe landing strategy of *Quetzalcoatlus* with 5 m/s headwind. Without headwind, landing might be hazardous to *Quetzalcoatlus*, because the safe landing speed is about 5 m/s, somewhat lower than the calculated speed of 7.3 m/s.

enabling the tailless pterodactyloids to steer while flying (Marsh, 1882; Heptonstall 1971). Bramwell and Whitfield (1974) concluded from wind tunnel study that the crest in *Pteranodon longiceps* was primarily a weight-saving device to reduce the size and weight of the neck musculature. It could serve as an airbrake by turning the head sidewise during landing. Stein (1975) suggested that the crest probably functioned as an attachment structure for a vertical membrane that was useful during turning and feeding.

Sexual Dimorphism

One of the reproductive strategies on size in animals is sexual dimorphism, showing a difference in appearance between the males and the females. In many birds, the males show elaborate and fantastic plumage. Bennett (1987, 2003) attributed the cranial crest of pterosaurs to be a male mating display feature, like the horns and antlers of mammals, for attracting females. He recognized sexual dimorphism among pterosaurs on the basis of the development of the crest. In male pterosaurs, recognized from the large body size and narrow pelvic cavity, the crest is enormous, whereas in females, the crest is small and the pelvic canal is wide to pass eggs. The crest was probably colorful in life and was possibly used for species recognition, allowing members of the same species to interact in group activities, such as feeding and flight. The crest was developed late in ontogeny during sexual maturity. Although the immensity of some male crests seems like a handicap during flight, when it comes to mating and display, it is clear that size matters. The American white pelican (*Pelicanus erythrorhynchos*) has a similar fibrous epidermal crest on top of the skull, toward the tip, as a display structure that grows in breeding season but is much reduced at other times (Sibley, 2001).

Of course, there are other pterosaurs, especially "rhamphorhynchoids," that completely lack the sagittal crest, and the sexual dimorphism is not so pronounced. Recently, Czerkas and Ji (2002) described an exquisitely preserved "rhamphorhynchoid" pterosaur, *Pterorhynchus,* from the Late Jurassic Haifangou Formation of China, with an unusual headcrest and a hairlike body covering. The crest is represented by both a small-ossified part, and a soft keratinous tissue element that is much larger and higher, extending over two-thirds of the skull. These authors concluded that the headcrest of *Pterorhynchus* might have acted as a rudder during flight.

Although the size, number, and position of headcrests are variable, all pterodactyloids had a proportionately enormous skull that was longer than the body length. How was the head oriented and what was the function of this enormous head during flight? The inner ear structure, especially the canalicular system revealed from the X-ray computerized tomography (CT) scan of the endocranial cavity, provides a clue to the habitual orientation and posture of the pterosaur head (Witmer et al., 2003). In tetrapods, the osseous labyrinth or three semicircular canals are oriented in space (two vertical and one horizontal) as a balancing organ, which sense the movements of the head. The preferred head orientation involves maintaining the horizontal semicircular canal approximately level with the horizon. Orientation of the head posture between "rhamphorhynchoids" and pterodactyloids relative to the long axis of the skull was quite different. Using this horizontal canal as a guide, we found that the long axis of the skull of *Anhanguera* was strongly down-turned, about 40° as in pelicans, but the neck was held in a sub-horizontal and more stable position (Witmer et al., 2003). Bramwell and Whitfield (1974) suggested that the crest in *Pteranodon* could reduce the size and weight of neck musculature. However, if the head was tilted toward downward in front of the atlanto-occipital joint it would not permit reduction in neck musculature. This tilted head posture would allow a less obstructed view from the crests when using binocular vision during flight or feeding (Figs. 35D, 35E). We estimated that the increase of horizontal drag due to downward tilt of the head of *Anhanguera* at a flight speed of 10 m/s

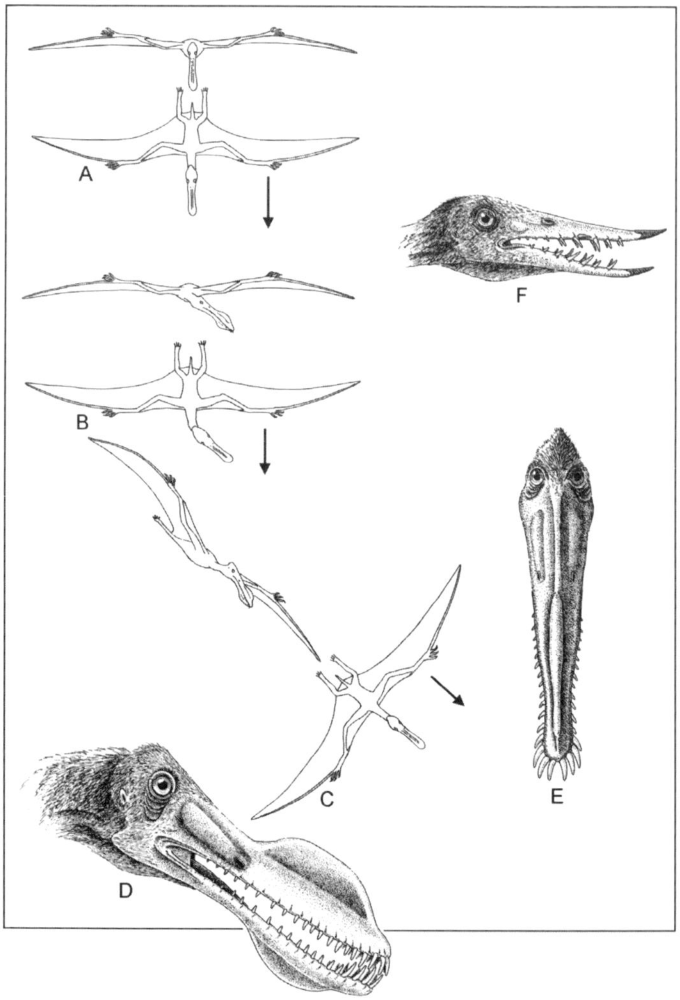

Figure 35. A–C: The head as a possible flight-control mechanism for large pterodactyloids such as *Anhanguera.* D–E: The head posture of *Anhanguera* at 40° downturned during flight, F: The head posture of "rhamphorhynchoids" during flight was horizontal (D–F, simplified from Witmer et al., 2003).

is negligible and would not interfere with its flight performance. In contrast, "rhamphorhynchoids" normally oriented their heads horizontally as revealed from the orientation of the horizontal canal (Fig. 35F).

It is likely that pterosaurs had gular pouches like modern pelicans to store fish and for sexual display. Impressions of the gular pouch can be seen in one of the *Pterodactylus* specimens (Frey and Martill, 1998, fig. 3; Wellnhofer, 1991a, p. 161). Living pelicans use their pouches in breeding displays that involve lifting the head to show off the pouch or inflating with air (Sibley, 2001).

Thermoregulation

Bone histology suggests that pterosaurs grew rapidly to adult size like modern birds, an indication of warm-bloodedness (Ricqles et al., 2000); they grew fast and bred early in life. It is likely that pterosaurs were diurnal where the flight membrane might have acted as a reflector of solar heat (Tischlinger and Frey, 2002). The exquisitely preserved skin membrane of pterosaurs with complex, multilayered structure, and blood vessels indicate how the wings may have been used in assisting the animals' cooling mechanism of body heat (Martill and Unwin, 1989; Tischlinger and Frey, 2002). In addition to the wing membrane, the cranial crests of pterosaurs might have acted as an additional thermoregulator organ. Recently, Kellner and Campos (2002) discovered a complex network of vascular channels on the bony surface of the enormous cranial crest in the Early Cretaceous pterosaur *Thalassodromeus* from Brazil. They concluded that the crest, irrigated by blood vessels, might have been used for thermoregulation, enabling the animal to shed body heat (Fig. 36). While resting with the crest facing the sun, the capillaries, filled with blood and circulating beneath the skin, could have absorbed additional heat. While flying, the crest could have lost the excess body heat, which would have been created by strenuous energetic activity.

Aerial Turns

The primary function of the enormous head of pterodactyloids appears to be associated with a yawing or steering mechanism (Bramwell and Whitfield, 1974; Stein, 1975; McGowan, 1991, 1994). Gliding birds turn by twisting their wings lengthwise asymmetrically so that the trailing edge of the one wing is down while the trailing edge of the opposite wing is up to roll into a bank. Because of the presence of rigid actinofibrils, the wing-twisting system might not have been very effective for pterosaurs for aerial turns. Instead, they might have used their enormous heads as roll-control systems. By turning its head in the direction it desired to fly and by partial retraction of one wing by the femur in the same direction, a pterosaur could automatically turn the rest of its body and change the course of its flight (Fig. 35A–C). The slow flying speeds would have conferred tight turning, which would allow pterosaurs to exploit narrow thermals (Bramwell and Whitfield, 1974).

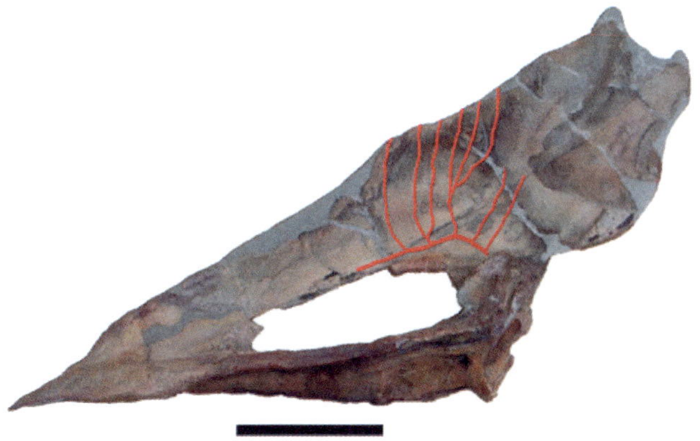

Figure 36. The skull of *Thalassodromeus sethi* from the Early Cretaceous Santana Formation of Brazil with an enormous crest showing channels of blood vessels (cast, TTU P10397); scale bar 15 cm; (modified from Kellner and Campos, 2002).

ECOLOGY, EVOLUTION, AND EXTINCTION

Pterosaurs were a successful group of flying archosaurs that appeared at the dawn of the age of dinosaurs in the Late Triassic some 225 million years ago. They dominated the sky for the next 160 million years throughout the Mesozoic and then suddenly disappeared at the end of Cretaceous. They colonized all continents and evolved a vast array of shapes and sizes. The central theme of pterosaur evolution was flight, and to maintain efficiency in the air, they minimized their weight as much as possible. They had very lightly built skeletons with air spaces in many of the bones and often with internal struts to provide structural integrity; they were natural marvels of flight engineering and structure. In critical regions of pterosaur skeletons—the thorax, outer wings, pelvis, and hindlimbs—a number of bones that occur separately in nonflying archosaurs were reduced, fused together, or modified, conferring enhanced structural rigidity that would help pterosaurs endure the enormous mechanical stresses of flying. The pterosaur skeletons combined lightness and strength, and in all their parts form beautifully followed function.

In spite of their delicate bones, there are several examples of immaculate preservation of pterosaurs from the Late Jurassic deposits (Solnhofen Limestone of West Germany, the Karabastau Formation of Kazakhstan in Central Asia, and the Chaomidianzi Formation of China), Early Cretaceous deposits (the Santana Formation of Brazil and the Tugulu Formation of China), and from the Late Cretaceous Niobrara Formation of Kansas. Many specimens from these localities afford a wealth of anatomical information, including skin, wing membranes, hair coverings, and stomach contents. The fossil record of pterosaurs provides a rich supply of morphological and aerodynamic data, but the origin and phylogeny of pterosaurs are still poorly understood.

Ecology

The predominant occurrence of the known pterosaur fossils in nearshore marine rocks suggests that they lived by the sea and islands and fed on marine organisms. Like many seabirds, smaller pterosaurs were probably inshore species, lived on islands and coastal areas, and roosted in trees or on cliffs, scrambling about like seabirds when not in the air (Fig. 37). They may be imagined as generally sleeping on clifftops and finding food on shore or no further at sea than 5 or 10 km. They might commute to and from feeding grounds when incubating or when bringing food for the young. They could take off from a perch simply by spreading their wings, and they could land and take off from water.

Pterosaurs developed particularly efficient sensory systems, such as visual and auditory acuity for the coordination of flight activity, as revealed from the study of endocasts (Jerison, 1973; Witmer et al., 2003). The brain seems to have filled the cranial cavity almost completely like that of modern birds, where the enlargement of cerebrum has led to the contact with the cerebellum dorsally, thus displacing the large optic lobes ventrally and laterally. The well-developed optic lobes and large, frontally placed eye sockets indicate that pterosaurs had excellent binocular vision for flight and feeding. As the small olfactory lobes hint, the sense of smell is relatively poorly developed in pterosaurs. The inner ear is also well developed, consisting of three enlarged semicircular canals and elongated lagena for balance and auditory acuity, possibly to supplement the sense of smell.

Pterosaurs might have spent most of day feeding in water. Large pterodactyloids such as *Anhanguera* and *Pteranodon* were most likely pelagic like albatrosses, spending much of their time on the wing, throughout the vast stretch of oceans, which would provide them uniform habitat, abundant food resources, and ample wind power to assist flight. Some pterosaurs, such as *Dsungaripterus* and *Quetzalcoatlus* were possibly migratory pterosaurs, oscillating between land feeding and shore-feeding. The fossil record indicates that they ventured inland and congregated near large lakes and rivers for foraging. The toothless tips of the beaks of *Dsungaripterus* are slightly bent upwards and were probably used to crack open bivalves, snails, and crabs like some extant shorebirds (Young, 1973). Because the fossils of *Quetzalcoatlus* have been found 400 km away from the coastline, it has been suggested that they might have scavenged dinosaur carcasses like a colossal vulture (Lawson, 1975). Langston (1981) preferred an invertebrate diet for *Quetzalcoatlus,* which probably employed its slender beak to probe mollusks or arthropods living in shallow flood basins. More likely, large pterodactyloids (M > 4.5 kg) such as *Dsungaripterus, Anhanguera, Pteranodon*, and *Quetzalcoatlus* were waders or surface riders during feeding, folding their wings sidewise and dipping their heads at 40° downward to feed on schools of fish (Fig. 37) like modern pelicans do, and they might have similar gular pouches for storing fish (Wellnhofer, 1991a); their long, narrow, and edentulous beaks could have been used like chopsticks to catch not only fish, but also crustaceans and other invertebrates while submerging on the shallow water. Lockley et al. (1995) described some pterosaur tracks dominated by the manus impressions and beak marks. They postulated that the trackmaker was buoyant and feeding on shallow water, supporting the body weight mainly on the forelimbs (Figs. 9B, 37). Their long and horizontal necks would have been adapted to scan large areas for foraging (Fig. 37). The vertebral column and the hindlimbs of the animals would remain buoyant on the water surface in horizontal fashion without muscular effort. On the water, pterosaurs would swim like geese and swans using their feet to propel themselves while folding their wings sidewise. During swimming they could raise their wings vertically upward like a sailboat using wind energy to travel large territory for foraging.

Seabirds show many styles of fishing, ranging from fish pirating in the air by frigatebirds (kleptoparasitism) to skimming, plunge diving, and underwater pursuit, depending on the size of the animal. The skeletal design of pterosaurs does not suggest a diving adaptation (Brower, 1983). Similar to brown pelicans, many small and medium-sized pterosaurs (M < 8 kg) with good flying ability could forage by plunge diving from the air into the water. They would begin a steep dive after sighting prey and enter water with wings stretched back. Witmer et al. (2003) argued that the large focculus in pterosaurs, connected with the vestibular system, the eye muscles, and the neck muscles, would permit a rock-steady gaze as they pursued their prey. Similarly, the head-down position in pterodactyloids might be advantageous for aerial fish cachers for two reasons. First, it would have enabled them to retain the prey item within the field of vision until they were ready to strike, and second, in this position, the jaws with the fish grab dentition were already deployed in a position suitable for snatching their prey from the water. On contact with water, the pterosaurs would fold their wings and open their bills, centering the prey between their jaws while sitting on the water's surface. Kellner and Langston (1996) suggested that *Quetzalcoatlus* might have caught fish like a skimmer bird while flying low over water with its lower jaw partially submerged. Later, Kellner and Campos (2002) extended this skimmer model for *Thalassodromeus*, another large pterodactyl from Brazil. Although skimming adaptation might be possible for small pterosaurs ($M < 0.3$ kg) while flying low over water, we doubt this skimming capability existed for large pterodactyloids because they were less maneuverable and might not maintain flight against water resistance. Moreover, skimmers do not have pointed beaks, as shown in *Dsungaripterus, Anhanguera*, *Pteranodon*, *Quetzalcoatlus*, and *Thalassodromeus*. They have a rather blunt jaw tip (which large pterodactyloids lack) to direct the water away to either side of the jaw as they skim.

Pterosaurs had a fine covering of hair over their body, indicating that they were warm-blooded (Fig. 36). This would provide the high metabolic rate that was necessary for an active flying lifestyle. Endothermic physiology demanded a large amount of food. Apparently, pterosaurs spent a great deal of time eating. The enormous skull and narrow snout facilitated the large intake of food. Fossilized stomach contents of many pterosaurs indicate that fish was one of their preferred diets because of its abundance in sea, but how the fish were caught is not clear. A wide variety of tooth mor-

Figure 37. Paleoecology of pterodactyloids such as *Anhanguera* in shallow marine environments where downturned head was used habitually for catching fish, or during flight like modern pelicans; manual trackmarks suggest that some pterodactyloids were buoyant and feeding on shallow water, supporting the body weight mainly on the forelimbs.

phology and loss of teeth in some forms indicate that pterosaurs had different styles of feeding habit (Wellnhofer, 1991a). Sharp fangs and large jaw muscles are the characteristic features of early pterosaurs as prey-catching devices. The neck was also highly mobile and constructed in an S-shaped fashion to deliver rapid lunges to grab at prey. The teeth, where they are present, are long and fang-like, adapted for catching slippery fish. Fish and other seafood must have been an important dietary component of many pterosaurs. Some large pterodactyloids are entirely toothless, like birds, and probably had horny beaks. Some used their numerous small and slender teeth for filtering small aquatic organisms. Others show flexible, bristle-like teeth on their lower jaws for sieving plankton. Some fed on insects, others on fruits and seeds. Some teeth were adapted for shell crushing. Other used their wide bill to skim food from the water's surface. Pterosaurs show considerable variety in the jaw apparatus for procuring different kinds of food.

Origin and Evolution

Archosaurs were the most prolific groups of Mesozoic reptiles that split into two clades—Pseudosuchia (= Crurotarsi) and Ornithodira—during the Triassic. The former lineage led to crocodilians via various extinct groups; the latter to pterosaurs, dinosaurs, and birds (Gauthier, 1986; Sereno, 1991). Although pterosaurs are currently classified as the basal major group of ornithodiran archosaurs mainly on the basis of hindlimb morphology (Fig. 1), their origin, evolution, and phylogenetic relationships are still obscured. *Scleromochlus* from the Late Triassic Lossiemouth Formation of Scotland, a small (180 mm long from snout to tail tip) archosaur, has been generally considered to be most closely related, or ancestral to, Pterosauria (Padian, 1983, 1985; Gauthier, 1986), but others have questioned its close phylogenetic proximity to pterosaurs (Sereno, 1991; Wellnhofer, 1991a; Benton, 1999). *Scleromochlus* is difficult to interpret because all tiny skeletons are preserved as natural molds in coarse sandstone matrix where many details are unavailable. One of the unusual features of *Scleromochlus* is the strong limb disparity where the hindlimb is much longer than the forelimb, indicating that the animal was a bipedal cursor with a digitigrade stance. In pterosaur ancestry, we would expect the opposite trend in limb proportion, where the forelimbs would be more elongated to assume the function of the protowing; similarly the pes would be plantigrade as seen in pterosaurs. In both accounts, *Scleromochlus* appears to be too derived for pterosaur ancestry, but closer to theropod lineage. Several osteological features in the manus and pes (e.g., recurved claws with flexor tubercles and elongated penultimate phalanges) indicate the scansorial ability of early pterosaurs; similarly the attachment of patagium to the hindlimb suggests that primitive pterosaurs were quadrupedal, arboreal forms (Unwin, 1988a; Bennett, 1997b). In contrast, *Scleromochlus* does not show any arboreal adaptation. Both phylogeny and functional anatomy indicate that *Scleromochlus* appears to be more derived than pterosaurs but closer to dinosaurs. Recently, Atanassov (2002) reported in his dissertation two ornithodiran archosaurs from the Late Triassic Dockum Group of Texas that may be a close sister-group to pterosaurs. Contrary to the traditional view that pterosaurs were the major sister-group of dinosaurs, Bennett (1996) proposed a radical hypothesis on the basis of detailed cladistic analysis that pterosaurs were primitive groups of basal archosauromorphs close to erythrosuchids. He argued that many archosaurian features of pterosaurs might have evolved convergently because of their functional similarities with the bipedal locomotion of theropods. Similarly, Peters (2000) proposed that *Sharovipteryx*, an enigmatic prolacertiform from the Late Triassic Kirghizia, with well-developed uropatagia, may be sister taxon to the Pterosauria.

Although no direct ancestor of pterosaurs is known in the fossil record that can bridge the crucial morphological gap, Wild (1983) reconstructed a hypothetical ancestor of pterosaur (protopterosaur) as a small, tree-climbing animal, which appears to be logical in form and function from morphological and ecological constraints. In this animal, there was a tendency to enlarge the body surface by a flap of skin, which extended from the caudal margin of the forelimb and the fourth digit of the hand to the leg like some modern gliding animals. It is possible that pterosaurs evolved from a similar arboreal protopterosaur, as Wild envisioned, that hopped from branch to branch, steadying themselves with outstretched forelimbs. As patagium enlarged on the margins of the forelimbs, the hops were likely extended. The animals may have leaped from branch to branch or tree to tree similar to modern leaping primates, steadying themselves with extended wings, and controlling their jumps and falls (Bennett, 1997b). Eventually, they began to glide using gravity. As gliding became specialized, active flight would evolve gradually. The arboreal model of pterosaur flight, whether parachuting (Wild, 1983; Pennycuick, 1986; Unwin, 1988a) or leaping (Bennett, 1997b), seems more parsimonious than the cursorial model that maintains that pterosaur flight evolved in running bipeds through a series of short jumps into the air without any gliding stage (Padian, 1983, 1985). The attachment of patagium to the hindlimb argued against the cursorial theory.

Body Size

There are about 120 species of pterosaurs known to date, ranging in size from as small as a sparrow to as large as a fighter plane, but their interrelationships are poorly known. From "rhamphorhynchoids" to pterodactyloids, there are several evolutionary trends, such as increasing body size, crest development, confluence of naris with antorbital opening, digital reduction, truncation of tail, downturned head, and loss of teeth, but the lack of detailed phylogenetic information precludes to trace these tends in a comprehensive way. Like many groups of animals, there is an apparent trend of body size increase during the evolutionary history of pterosaurs. During the Late Triassic and Jurassic, pterosaurs were small to medium-sized, but they became progressively larger in the Cretaceous. During the Late Cretaceous, many of the last pterosaurs, such as *Pteranodon* and *Quetzalcoatlus* were gigantic soarers, breaking the size limits of previously known flying animals. This evolutionary trend towards large body size in many lineages was first formalized by Cope (1887), who documented that the average size of mammals increased dramatically during the Cenozoic. He attributed this pattern to a tendency for new groups to evolve at small sizes, combined with an innate drive toward larger size. This has been traditionally codified as Cope's Rule, defined as evolutionary increases of body size over time. The apparent body size increase in several lineages of vertebrates, such as pelycosaurs, archosaurs, horses, camels, proboscideans, and whales, is well documented as an example of anagenetic evolution (Stanley, 1973; McKinney, 1990). The reason seems to be that most groups arose at small size relative to their potential size ranges. We do not know what selective forces produced these repeated trends, although it is assumed that they are direct result of natural selection. There are many exceptions, however, to this trend. Cope's Rule may be true from certain perspectives but is no longer generally accepted. The advantages of increasing body sizes in a clade are certainly not the progressive internal drive of the animal (known as orthogenesis), as Cope speculated, but may be due to some ecological forces. Some of the potential advantages to large size are greater efficiency of predation, predator deterrence, heat retention, increased intelligence, greater reproductive success, extended individual mortality, and exhibiting "K selection" with expanded size range of possible food items (reviewed in Stanley, 1973). Size change can also stem from ecophenotypic reasons, such as abundant food, which can produce large-sized animals, while less providence causes stunted growth. There are potential disadvantages for being large, such as greater demand on resources, more restricted nature of niches, small population sizes, and restricted gene pools. A large animal requires a large niche, and is, therefore, automatically limited. These factors promote higher rates of extinction for larger animals during a mass extinction event. Smaller is better, for example, when it is advantageous to reproduce early and often, exhibiting "r selection." Ecosystems afford more niches for small species than large ones. K selection and r selection are the two extremes of a range of evolutionary strategies for pterosaurs for reproductive increase. Early pterosaurs with small size showed preference for r selection, where production of large numbers of offspring was insurance against juvenile mortality or environmental catastrophe. Later pterosaurs exhibited K selection, with large size and long generation of time where the full resources of environments were exploited safely.

Gould (1996) interpreted Cope's Rule in reverse: he suggested that evolutionary size increase is largely a matter of cladogenetic diffusion away from originally small-sized ancestors and hence the only way to evolve is up and toward specialization. He presented a pedagogical metaphor called the "Drunkard's Walk" to explain the apparent increase in size within a phyletic lineage. This model is based on the random walk of a staggering drunk man, with a wall on the left side and a gutter on the right. The average path taken will tend to move away from the solid wall towards the gutter where there is no obstruction. If the drunkard staggers long enough, he will eventually end up in the gutter. In other words, the only direction of movement that remains open to him when constrained by a left wall is toward the gutter, but his motion indicates no trend. Gould compared the random walk to a taxonomic lineage with the wall as the lower size limit. Founding species start at the left wall, and the range of size can therefore expand in only one direction toward the larger size. Thus, Cope's Rule, according to Gould, identifies a predominant relative frequency, not an absolute trend.

Superficially, the effect of Cope's Rule appears to be stronger for pterosaurs than for Mesozoic birds. Pterosaurs arose with birds during the Late Triassic and shared the skies for 160 million years. During this period the average body mass of pterosaurs increased dramatically, almost 4700 times, from the small Triassic form such as *Eudimorphodon*. In contrast, Mesozoic birds do not show such a clear-cut trend. During this same span, the body mass of Mesozoic birds increased merely 100 times from the Triassic bird *Protoavis* to the largest flightless diving forms such as hesperornithiforms, but the mass of the volant birds had a narrow range from 0.2 kg to 1 kg (Chatterjee, 1997). However, when the body mass of selected taxa of pterosaurs is plotted against phylogeny spanning millions of years (Fig. 38), we see erratic irregularities of increment of body size like the fluctuations of market price of many stocks over years. In recent times, Brownian motion, the jittering of small particles suspended in liquid, has been applied to economics to make long-term predictions. For example, one-dimensional Brownian motion has been found to correspond roughly to the fluctuations of the stock market (Chance and Peterson, 1999). The study of major trends in body-size in phylogenetic context is intriguing and complex, like the rise and fall of the stock market. Perhaps, in the future, we will gain a better understanding of whether the equation of Brownian motion obeys Cope's Rule, when we have large data sets of body mass and refined phylogeny of pterosaurs.

Evolution has emancipated the last pterosaurs from the constraints of flying size. We must look for proximate and ultimate causes to study the phenomenon of increasing body size in pterosaurs. Larger animals tend to live longer. There are other ecological benefits for large animals, as discussed earlier, but there are some penalties too. The large size of pterosaurs may be linked to the physics of heat transference as a function of the surface-to-volume ratio. It has been known for more than a century that as animals get larger their body masses (M) increase approximately with the cube of their length, whereas their surface areas increase only with their square. Because the heat an animal can shed is proportional to skin surface, its maximum continuous metabolic rate is proportional to surface area, which is itself proportional

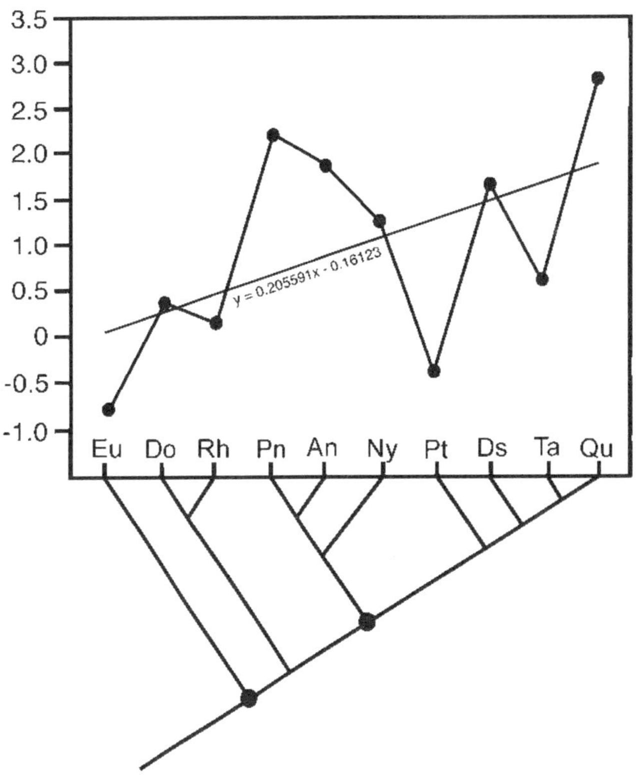

Figure 38. Trends in body-size evolution in pterosaurs. Log of body weight (in Newton) of ten pterosaur genera, when plotted against their phylogeny, shows episodic fluctuations of body size like the rise and fall of stock markets; yet the long-term trend, as shown by the regression line indicates increase in size during the span of 160 million years.

to $M^{2/3}$. This implies that the metabolic rate per kg of mass decreases with increasing size, inversely as the cube root of mass, though some recent analysis suggests that the correct scaling law is nearer to the fourth root. As animals become larger, their metabolic rate slows down in proportion to the quarter power of an animal's body mass; the correlation between body size and total metabolic rate may stem from nutrient distribution in fractal-like vascular networks (West et al., 1999). In the case of pterosaurs, the aerodynamic power requirement for flapping flight would increase with body mass more rapidly than available metabolic power. Slow metabolism and lack of enough power for large pterodactyloids would preclude them from sustained flapping flight and would lead them to search for free rides in ascending currents of air. Possibly they had a small mass of flight muscles relative to their large body mass. Even then, they only needed enough muscle power during takeoff to become airborne; they would exploit winds or thermals to stay aloft. Thus, the increased large body size among Cretaceous pterodactyloids may indicate a dramatic switching of flight style from a sustained flight to a soaring flight; the latter had never been explored before. In contrast to pterosaurs, Mesozoic birds exploited a wide range of niches, showing four distinct ecological types: basal land birds, foot-propelled divers, shore birds, and flightless terrestrial birds (Chatterjee, 1997). Flightlessness is a recurrent theme in the evolution of birds. So far, there is no example of flightless pterosaurs. On the other hand, soaring flight style in birds evolved much later in the Early Tertiary by pseudodontorns, whose wings may have spanned as much as 7 m, thus filling the niches vacated by the Cretaceous pterosaurs.

The appearance of giant pterodactyloids on the Late Cretaceous scene is reminiscent of the evolution of the giant insects in the Carboniferous coal measures. Did any physical factor, such as atmospheric composition, contribute to the gigantism of the flying animal? Graham et al. (1997) argued that during the Carboniferous, the large amount of oxygen release by the swampy forests of the coal measures of Pangea altered the atmospheric composition. They suggest that oxygen levels were much higher during the Carboniferous than at any time before or since—it was 35%, compared with the 21% of the present level. In this oxygen-rich atmosphere of the coal measures, there was an evolutionary explosion of giant insects, including the dragonfly *Meganeura,* with its 70 cm wingspan. The hyperdense atmosphere would have provided more lift and made it easier for these insects to fly, according to Graham et al. (1997). More important, the excess oxygen would have diffused into their tissues more rapidly—thus enabling them to grow larger. In the Triassic, the oxygen level dropped dramatically (15%) below the present level but rose again steadily during the Jurassic and Cretaceous. A second but smaller elevation of oxygen concentration (~26%) occurred in the Late Cretaceous (Berner and Canfield, 1989) when large pterodactyloids such as *Pteranodon* and *Quetzalcoatlus* dominated the sky. Extensive Cretaceous coalfields were laid down at 50° paleolatitudes, indicating abundant release of oxygen into the atmosphere. Dudley (1998) correlated the gigantism of pterodactyloids with the oxygen-rich, dense atmosphere of the Late Cretaceous period. He argued that the physical effects of an increased air density would result in increased lift and possibly advantageous shifts of Reynolds numbers. We believe that an increased oxygen supply may enhance respiratory efficiency for these large pterodactyloids, enabling them to take off without switching to anaerobic physiology. However, increased air density would not have noticeable effect on flight L/D ratio, and therefore on drag, though flight speeds and required power might be slightly lower. The remark about the higher Reynolds number by Dudley (1998) is not relevant, since for the largest pterosaurs, the effects of further increasing *Re* on drag would have a negligible effect.

Extinction

There is no paleoecological or taphonomic evidence that suggests competition between pterosaurs and birds during the Mesozoic; in rare occasions, they are found together in the same sediments. Cretaceous pterodactyloids developed a great variety of forms. As they became larger, they became edentulous, where the beaks were covered with rhamphotheca. The Upper Cretaceous saw the last and arguably most exciting period of pterosaur evolution when three families of giant pterodactyloids were dominant in the sky: the Nyctosauridae and Pteranodontidae in North America and Europe, and the Azhdarchidae (which include *Quetzalcoatlus* with great elongated neck vertebrae) from North America, Europe, Central Asia, and Jordan. By the Maastrichtian, azhdarchids had achieved global dominance. Unwin (1988b) pointed out that during the Late Cretaceous, as bird diversity rose, that of the pterosaurs steadily declined. However, it was not competition with birds that led to the elimination and extinction of pterosaurs. It was an example of ecological replacement without competition, not a competitive displacement. The expansion and explosive evolution of birds in the Tertiary is merely opportunistic, following the demise of pterosaurs at the end of the Cretaceous (Chatterjee, 1997).

The Cretaceous-Tertiary (K-T) mass extinction wiped out the dinosaurs, pterosaurs, plesiosaurs, mosasaurs, and two-thirds of all marine species 65 million years ago. Having survived for 160 million years, the pterosaurs' sudden extinction along with dinosaurs at the end of the Cretaceous is puzzling. The large size of the last pterodactyloids during the biotic crisis may be one the factors for their demise, because all land animals more than 25 kg disappeared from the planet at the K-T extinction. A higher extinction rate for larger taxa may have resulted from both small population size and a high degree of specialization. Many environmental scenarios such as drop of global temperatures, increasing seasonal variation of weather, disappearance of warm, shallow seas, prevalence of stormy, gusty winds, lack of thermal updrafts, etc., have been evoked to explain the demise of pterosaurs (Langston, 1981), but none appears to be the proximate cause. The K-T extinction was multi-causal and more lethal in nature, which affected simultaneously both land and sea creatures. Currently, two competing models have been proposed to explain the apocalyptic disaster at the K-T boundary: volcanic hypothesis and meteoritic hypothesis. The volcanic theory argues that the pollution in the atmosphere and ocean by the massive outpourings of Deccan flood basalt in India had devastating effects on ecology (Courtillot, 1990). The impact theory postulates that the environments were lethally altered or destroyed at the end of the Cretaceous by collision of a large meteorite leading to biotic crisis (Alvarez et al., 1980). A new emerging point of view suggests that the K-T extinction may have been caused by two or more nearly synchronous impacts. In the recently proposed multiple impacts theory, a large asteroid broke up into a swarm of fragments and ricocheted downrange, excavating the Chicxulub crater in Mexico, Shiva crater in India, and DSDP 576 site on the North Pacific Ocean (Chatterjee and Rudra, 1996). Because of the close proximity of the Shiva crater and the Deccan Traps and analogy with lunar maria, it seems possible that the Shiva impact triggered the Deccan Trap volcanism, and both impact and volcanic catastrophes contributed heavily to the environmental disasters—including storms, acid rain, global pollution and wildfires, tsunamis, nuclear winters, and earthquakes—that all played major roles in the killing mechanism. The world's oceans became sufficiently acidic to dissolve calcareous shell material, thus collapsing the marine food chain and affecting the top carnivores. Since both the Chicxulub and Shiva impacts occurred at the coastal regions, huge tsunamis produced by the impact would have destroyed shallow-marine habitats across the globe, devastating the sanctuary for the Maastrichtian pterodactyloids. In contrast, contemporary birds were relatively small, were generalists, were adapted to a wide range of habitats in both terrestrial and marine realms, and had ecological superiority over pterosaurs during crisis (Chatterjee, 1997). Their small size and versatile lifestyles may have enhanced survival during the K-T mass extinction. Although several archaic groups of birds, such as hesperornithiforms, ichthyornithiforms, and enantiornithines, became victims at the K-T extinction, some modern groups, such as loons, presbyornithids, and charadriiforms, colonized in Antarctica during the Maastrichtian, transcended the K-T extinction, rebounded, and radiated explosively in the Tertiary (Chatterjee, 2002). The success of birds over pterosaurs during the Cretaceous crisis may be linked to differential survival strategy (Unwin, 1988b). Large-bodied pterosaurs had two disadvantages: they generally had smaller populations and lower reproductive rates than smaller-bodied bird species. When struck by a catastrophe, they were more prone to extinction. In contrast, birds in high latitudes endured mass extinction at the end of the Cretaceous because of their small size and wide ecological adaptations. They radiated explosively into the Tertiary period, spreading into ecological niches that the pterosaurs had vacated, and became the dominant kind of flying animals.

ACKNOWLEDGMENTS

We thank Peter Wellnhofer (Bayerische Staatssammlung für Päläontologie und Historische Geologie, Munich), Rupert Wild (Staatliches Museum für Naturkunde, Stuttgart), David Berman (Carnegie Museum of Natural History, Pittsburgh), John Maisey (American Museum of Natural History, New York), and Dong Zhiming (Institute of Vertebrate Paleontology and Paleoanthropology, Beijing) for access to pterosaur specimens in their collections. We thank Alex Kellner for providing exquisite, three-dimensional cast skeletons of *Anhanguera*, *Thalassodromeus*, and *Tapejara*, and Matt Smith for the cast skeleton of *Quetzalcoatlus*. We thank Richard Strauss and Momchil Atanassov for statistical analysis and estimation of mass and wing area of pterosaurs, and Kyle McQuilkin for animation of terrestrial locomotion. We especially thank Peter Wellnhofer for his *Illustrated Encyclopedia of Pterosaurs*, which contains valuable information on the pterosaurs of the world. We thank many pioneers who attempted to unravel the locomotion of pterosaurs. Momchil Atanassov, Larry Witmer, Kevin Padian, Richard Brower, James DeLaurier, and Chris Bennett offered valuable suggestions and reviewed the manuscript at several stages. We especially thank Kyle McQuilkin, Mike Nickell, Jeff Martz, and Momchil Atanassov for help in preparation of figures. The research was supported by Tübingen University and Texas Tech University.

REFERENCES CITED

Abel, O., 1925, On a skeleton of *Pterodactylus antiquus* from the lithographic shales of Bavaria, with remains of skin and musculature: American Museum Novitates, v. 192, p. 1–12.

Abbott, I.H., and Doenhoff, A.E. von, 1959, Theory of wing sections: New York, Dover Publications, 693 p.

Alexander, R.M., 1992, Exploring biomechanics: Animals in motion: New York, W.H. Freeman Co., 247 p.

Alvarez, L.W., Alvarez, W., Asaro, F., and Michel, H.V., 1980, Extraterrestrial cause for the Cretaceous-Tertiary extinction: Science, v. 208, p. 1095–1108.

Atanassov, M.N., 2002, Two new archosaur reptiles from the Late Triassic of Texas [Ph.D. dissertation]: Lubbock, Texas Tech University, p. 352.

Atanassov, M.N., and Strauss, R., 2002, How much did *Archaeopteryx* and *Quetzalcoatlus* weigh? Mass estimation by multivariate analysis of bone dimension [abs.]: Journal of Vertebrate Paleontology, v. 22, no. 3, p. 33A.

Avallone, E.A., and Baumeister, T., editors, 1987, Marks' Standard Book for Mechanical Engineers: New York, McGraw Hill, 1890 p.

Bartholomew, G.A., and Casey, T.M., 1978, Oxygen consumption of moths during rest, pre-flight warm-up and flight in relation to body size and wing morphology: Journal of Experimental Biology: v. 76, p. 11–15.

Bennett, S.C., 1987, New evidence on the tail of the pterosaur *Pteranodon* (Archosauria: Pterosauria), *in* Currie, P.J., and Koster, E.H., eds., Fourth Symposium on Mesozoic Terrestrial Ecosystems, Short Papers: Drumheller, p. 18–23.

Bennett, S.C., 1994, Taxonomy and systematics of the Late Cretaceous pterosaur *Pteranodon* (Pterosauria, Pterodactyloidea), Occasional Papers: Lawrence, Kansas, Natural History Museum, University of Kansas: v. 169. p. 1–70.

Bennett, S.C., 1996, The phylogenetic position of the Pterosauria within the Archosauromorpha: Zoological Journal of the Linnean Society: v. 118, p. 261–308.

Bennett, S.C., 1997a, Terrestrial locomotion of pterosaurs: a reconstruction based on *Pteraichnus* trackways: Journal of Vertebrate Paleontology, v. 17, p. 104–113.

Bennett, S.C., 1997b, The arboreal leaping theory of the origin of pterosaur flight: Historical Biology, v. 12, p. 265–290.

Bennett, S.C., 2000, Pterosaur flight: the role of actinofibrils in wing function: Historical Biology, v. 14, p. 255–284.

Bennett, S.C., 2001, The osteology and functional morphology of the Late Cretaceous pterosaur *Pteranodon*: Palaeontographica, v. A 260, p. 1–112.

Bennett, S.C., 2003, New crested specimens of the Late Cretaceous pterosaur *Nyctosaurus*: Paläontologische Zeitschrift, v. 77, p. 61–75.

Benton, M.J., 1999, *Scleromochlus taylori* and the origin of dinosaurs and pterosaurs: Philosophical Transactions of the Royal Society of London, v. B354, p. 1423–1446.

Berner, R.A., and Canfield, D.E., 1989, A new model for atmospheric oxygen over Phanerozoic time: American Journal of Science, v. 289, p. 333–361.
Bramwell, C.D., and Whitfield, G.R., 1974, Biomechanics of *Pteranodon*: Philosophical Transactions of the Royal Society of London, v. B267, p. 503–581.
Brower, J.C., 1980, Pterosaurs: how did they fly?: Episodes, v. 1980, p. 21–24.
Brower, J.C., 1983, The aerodynamics of *Pteranodon* and *Nyctosaurus*, two large pterosaurs from the Upper Cretaceous of Kansas: Journal of Vertebrate Paleontology, v. 3, p. 84–124.
Brower, J.C., and Veinus, J., 1981, Allometry in Pterosaurs: Lawrence, Kansas, University of Kansas Paleontological Contributions, v. 105, p. 1–32.
Burton, R., 1990, Bird flight: New York, Facts on File, 160 p.
Campbell, K.E., and Toni, E.P., 1983, Size and locomotion in teratorns (Aves: Teratornithidae), Auk: v. 100, p. 390–403.
Chance, D.M., and Peterson, P.P., 1999, The new science of finance: American Scientist, v. 87, p. 256–263.
Chatterjee, S., 1982, Phylogeny and classification of thecodontian reptiles: Nature, v. 295, p. 317–320.
Chatterjee, S., 1992, A remarkable pterosaur skull from the Early Cretaceous of Brazil [abs.]: Journal of Vertebrate Paleontology, v. 12, no. 3, p. 23A.
Chatterjee, S., 1997, The rise of birds: Baltimore, Johns Hopkins University Press, 312 p.
Chatterjee, S., 2002, The morphology and systematics of *Polarornis*, a Cretaceous loon (Aves: Gaviidae) from Antarctica, *in* Zhou, Z., and Zhang, F., eds., Proceedings of the 5th Symposium of the Society of Avian Paleontology and Evolution, Beijing: Science Press, p. 125–155.
Chatterjee, S., and Rudra, D.K., 1996, KT events in India: impact, rifting, volcanism and dinosaur extinction: Memoirs of the Queensland Museum, v. 39, p. 489–532.
Chatterjee, S., and Templin, R.J., 2001, The flight of pterosaurs [abs.]: Journal of Vertebrate Paleontology: v. 21, no. 3, p. 40A.
Chatterjee, S., and Templin, R.J., 2003, The flight of *Archaeopteryx*: Naturwissenschaften, v. 90, p. 27–32.
Clark, J.M., Hopson, J.A., Hernandez, R., Fastovsky, D.E., and Montellano, M., 1998, Foot posture in a primitive pterosaur: Nature, v. 391, p. 886–889.
Collini, C.A., 1784, Sur quelques Zoolithes du Cabinet d'Histoire Naturalle de S.A.S.E. Palatine et de Baviere, a Mannheim, Acta Academiae Theodoro-Palatine Mannheim, pars physica: v. 5, p. 58–103.
Cope, E.D., 1887, The origin of the fittest: New York, Appleton, 547 p.
Courtillot, V., 1990, A volcanic eruption: Scientific American, v. 263, no. 4, p. 85–92.
Cuvier, G., 1801, Reptile volant, *in* Extrait d'un ouvrage sur les espèces de quadrupèdes dont on a trouvé les ossemens dans l'intérieur de la terre: Proces de la Classe Scientifique, Mathematique et Physique de l'Institute Nationale (26 Brumaire, l'an 9): Journal de Physique, de Chime et d'Histoire Naturalle, v. 13, p. 253–267.
Czerkas, S.A., and Ji, Q., 2002, A new rhamphorhynchoid with a headcrest and complex integumentary structures: The Dinosaur Museum Journal, v. 1, p. 15–41.
DeLaurier, J., 1989, An experimental study of low-speed single-surface airfoils with faired leading edges: Proceedings of the Conference on Low Reynolds Number Aerodynamics, Notre Dame, Indiana: Berlin: Springer-Verlag, p. 169–173.
Dial, K.P., Biewenwer, A.A., Tobalske, B.W., and Warrick, D.R., 1997, Mechanical power output for bird flight: Nature, v. 390, p. 67–70.
Döderlein, L., 1929, Uber *Rhamphorhynchus* und sein Schwanzsegel: Sitzungsberichte der Bayerischen Akademie der Wissenschaften zu München, Mathematisch-Naturwissenschaftlichen Klasse, p. 1–46.
Dudley, R., 1998, Atmospheric oxygen, giant Paleozoic insects and the evolution of aerial locomotor performance: Journal of Experimental Biology, v. 201, p. 1043–1050.
Frey, E., and Martill, D.M., 1998, Soft tissue preservation in a specimen of *Perodactylus kochi* (Wagner) from the Upper Jurassic of Germany: Neues Jarbuch für Geologie und Paläontologie, Abhandlungen, v. 210, p. 421–441.
Frey, E., and Riess, J., 1981, A new reconstruction of the pterosaur wing: Neues Jarbuch für Geologie und Paläontologie, Abhandlungen, v. 161, p. 1–27.
Gatesy, S.M., 1990, Caudofemoral musculature and the evolution of theropod locomotion: Paleobiology, v. 16, p. 170–186.
Gauthier, J.A., 1986, Saurischian monophyly and the origin of birds: California Academy of Sciences Memoirs, v. 8, p. 1–55.
Graham, J.B., Aguilar, N., Dudley, R., and Gans, C., 1997, The late Paleozoic atmosphere and the ecological and evolutionary physiology of tetrapods, *in* Sumida, S.S., and Martin, K.L.M., eds., Amniote origins: completing the transition to land: New York, Academic Press, p. 141–167.
Gould, S.J., 1996, Full house: New York, Harmony Books, 244 p.
Hankin, E.H., and Watson, D.M.S., 1914, On the flight of pterodactyloids: Aeronautical Journal, v. 18, p. 324–335.
Hazlehurst, G.A., 1991, The morphometric and flight characteristics of the Pterosauria [Ph. D. dissertation]: Bristol, University of Bristol, 274 p.
Hazlehurst, G.A., and Rayner, J.M.V., 1992a, Flight characteristics of Triassic and Jurassic Pterosauria: an appraisal based on wing shape: *Paleobiology*, v. 18, p. 447–463.
Hazlehurst, G.A., and Rayner, J.M.V., 1992b, An unusual flight mechanism in the Pterosauria: Palaeontology, v. 35, p. 927–941.
Henderson, D.M., and Unwin, D.M., 1999, Mathematical and computational model of a walking pterosaur [abs]: Journal of Vertebrate Paleontology, v. 19, p. 50A.
Heptonstall, W.B., 1971, An analysis of the flight of the Cretaceous pterodactyl *Pteranodon longiceps* (Marsh): Scottish Journal of Geology, v. 7, p. 61–78.
Hoerner, S.F., 1992, Fluid dynamic drag: Published privately by the author, © 1992, Lisolotte A. Horner, 456 p.
Holst, E. von, 1957, Der Saurierflug: Paläontologische Zeitschrift, v. 31, p. 15–22.
Hotton, N., III 1980, An alternative to dinosaur endothermy: the happy wanderers, *in* Thomas, R.D.K., and Olson, E.C., eds., A cold look at the warm-blooded dinosaurs: Boulder, Colorado, Westview Press, American Association for the Advancement of Science, p. 311–350.
Jenkins, F.A., Jr., Dial, K.P., and Goslow, G.E., Jr., 1988, A cineradiographic analysis of bird flight: the wishbone in starlings is a spring: Science, v. 241, p. 1495–1498.
Jerison, H.J., 1973, Evolution of the brain and intelligence: New York, Academic Press, 482 p.
Kardong, K.V., 2002, Vertebrates: comparative anatomy, function, evolution: New York, McGraw Hill, 761 p.
Kauffman, J., 1970, Birds in flight: New York, William Morrow, 96 p.
Kellner, A.W.A., 1989, A new edentate pterosaur of the Lower Cretaceous from the Araripe Basin, Northeast Brazil: Anais da Academia Brasileira de Ciências, v. 61, p. 339–446.
Kellner, A.W.A., 1995, The relationships of Tapejaridae (Pterodactyloidea) with comments on pterosaur phylogeny, *in* Sun, A., and Wang, Y., eds., Sixth Symposium on Mesozoic Terrestrial Ecosystems and Biota, Short papers: Beijing, China Ocean Press, p. 73–77.
Kellner, A.W.A., and Campos, D.A., 2002, The function of the cranial crest and jaws of a unique pterosaur from the Early Cretaceous of Brazil: Science, v. 297, p. 389–392.
Kellner, A.W.A., and Langston, W., Jr., 1996, Cranial remains of *Quetzalcoatlus* (Pterosauria, Azhdarchidae) from Late Cretaceous sediments of Big Bend National Park, Texas: Journal of Vertebrate Paleontology, v. 16, p. 222–231.
Kellner, A.W.A., and Tomida, A.Y., 1993, The ankle structure in a juvenile anhanguerid (Pterosauria, Pterodactyloidea) from the Lower Cretaceous of Brazil [abs.]: Journal of Vertebrate Paleontology, v. 13, no. 3, p. 44A.
Kellner, A.W.A., and Tomida, Y., 2000, Description of a new species of Anhangueridae (Pterodactyloidea) with comments on the pterosaur fauna from the Santana Formation (Aptian-Albian), northeastern Brazil: National Science Museum Monographs, v. 17, p. 1–135.
Kreider, J.F.K., 1985, Principles of fluid mechanics: Boston, Allyn and Bacon, 367 p.
Kripp, D. von, 1943, Ein Lebensbild von *Pteranodon longiceps* auf flugtechnischer Grundlage: Nova Acta Leopoldina, v. 12, p. 215–246.
Langston, W., Jr., 1981, Pterosaurs, Scientific American: v. 244, p. 122–126.
Lawson, D.A., 1975, Pterosaur from the latest Cretaceous of West Texas: discovery of the largest flying creature: Science, v.187, p. 947–948.
Lighthill, M.J., 1975, Aerodynamic aspects of animal flight, *in* Wu, T.Y.T., Brokaw, C.J., and Brennen, C., eds., Swimming and flying in nature: New York, Plenum Press, p. 423–491.
Lockley, M.G., Logue, T.J., Moratalla, J.J., Hunt, A.P.P., Schultz, J., and Robinson, J.M., 1995, The fossil trackway *Pteraichnus* is pterosaurian, not crocodilian: implications for the global distribution of pterosaur tracks: Ichnos, v. 4, p. 7–20.
MacCready, P., 1985, The great pterodactyl project: Engineering and Science, v. 49, p. 18–24.
Marden, J.H., 1994, From damselflies to pterosaurs: how burst and sustainable flight performance scale with size: American Journal of Physiology, v. 35, p. R1077–R1084.
Marsh, O.C., 1882, The wings of pterodactyloids: American Journal of Science, v. 23, p. 251–256.

Martill, D.M., and Unwin, D.M., 1989, Exceptionally well preserved pterosaur wing membrane from the Cretaceous of Brazil: Nature, v. 340, p. 138–140.

Mazin, J.M., Hantzpergue, P., Lafaurie, G., and Vignaud, P., 1995, Des pistes de pterosaures dans le Tithonien de Crayssac (Quercy, France): Comptes Rendus de l'Académie des Sciences, Série II, Mécanique-Physique-Chimie, Sciences de la Terre et de l'Univers: v. 321, p. 417–424.

McFarland, W.N., Pough, F.H., Cade, T.J., and Heiser, J.B., 1979, Vertebrate life: New York, MacMillan Publishing, 875 p.

McGowan, C., 1991, Dinosaurs, spitfires, and sea dragons: Cambridge, Harvard University Press, 365 p.

McGowan, C., 1994, Diatoms to dinosaurs: the size of living things: Washington, D.C., Island Press, 288 p.

McKinney, M.L., 1990, Trends in body-size evolution, *in* M.L. McKinney, ed., Evolutionary trends: Tucson, University of Arizona Press, p. 75–118.

McMahon, T.A., 1984, Muscles, reflexes, and locomotion: Princeton, Princeton University Press, 331 p.

McMahon, T.A., and Bonner, J.T., 1983, On size and life: New York, Scientific American Books, 255 p.

McQuilkin, K.S., Gedeon, A.T., and Dyer, C.T., 2002, Rearticulating exhibition mounts: new perspectives and opportunities; a case study on *Anhanguera* [abs.]: Journal of Vertebrate Paleontology: v. 22, no. 3, p. 86A.

Norberg, U.M., 1985, Flying, gliding, soaring, *in* Hildebrand, M., Bramble, D.M.K., Liem, F., and Wake, D.B., eds., Functional vertebrate morphology: Cambridge, Harvard University Press, p. 129–158.

Padian, K., 1979, The wings of pterosaurs: a new look: Discovery, v. 14, p. 20–29.

Padian, K., 1983, A functional analysis of flying and walking in pterosaurs: Paleobiology, v. 9, p. 218–239.

Padian, K., 1985, The origins and aerodynamics of flight in extinct vertebrates: Palaeontology, v. 28, p. 413–433.

Padian, K., 1988, The flight of pterosaurs: Natural History, v. 97, p. 58–65.

Padian, K., and Rayner, J.M.V., 1993, The wings of pterosaurs: American Journal of Science, v. 293A, p. 91–166.

Padian, K., and Wild, R., 1992, Studies of Liassic pterosauria, I. The holotype and referred specimens of the Liassic pterosaur *Dorygnathus banthensis* (Theodori) in the Petrefaktensammlung Banz, northern Bavaria: Palaeontographica, v. A225, p. 59–77.

Padian, K., and Olsen, P.E., 1984, The fossil trackway *Pteraichnus*: not pterosaurian, but crocodilian: Journal of Paleontology, v. 58, p. 178–184.

Paul, G.S., 2002, Dinosaurs of the air: Baltimore, Johns Hopkins University Press, p. 460.

Pennycuick, C.J., 1968, A wind-tunnel study of gliding flight in the pigeon *Columba livia*: Journal of Experimental Biology, v. 49, p. 509–526.

Pennycuick, C.J., 1972, Soaring behaviour and performance of some East African birds, observed from a motor glider: Ibis, v. 114, p. 178–218.

Pennycuick, C.J., 1982, The flight of petrels and albatrosses (Procellariformes), observed in South Georgia and its vicinity: Philosophical Transactions of the Royal Society of London, v. B300, p. 75–106.

Pennycuick, C.J., 1983, Thermal soaring compared in three dissimilar tropical bird species, *Fregata magnificens*, *Pelicanus occidentalis* and *Coragyps atratus*: Journal of Experimental Biology, v. 102, p. 307–325.

Pennycuick, C.J., 1986, Mechanical constraints on the evolution of flight: Memoirs of the California Academy of Sciences, v. 8, p. 83–98.

Pennycuick, C.J., 1988, On the reconstruction of pterosaurs and their manner of flight, with notes on vortex wakes: Biological Reviews of the Cambridge Philosophical Society, v. 63, p. 299–331.

Perkins, C.D., and Hage, P.E., 1949, Airplane perfor mance stability and control: New York, John Wiley & Sons, 403 p.

Peters, D., 2000, A reexamination of the four prolacertiformes with implications of pterosaur phylogeny: Rivista Italiana di Paleontologia e stratigrafic, v. 106, p. 293–336.

Peters, D., 2001, A new model for the evolution of the pterosaur wing—with a twist: Historical Biology, v. 15, p. 277–301.

Peters, R.N., 1983, The ecological implications of body size: Cambridge: Cambridge University Press, 329 p.

Poore, S.O., Ashcroft, A., Sanchez-Haiman, A., and Goslow, G.E., Jr., 1997, The contractile properties of the M. supracoracoideus in the pigeon and starling: a case for long-axis rotation of the humerus: Journal of Experimental Biology, v. 200, p. 2987–3002.

Prandtl, L., and Tietjens, O.G., 1934, Applied hydro- and aeromechanics: New York, Dover Publications, 312 p.

Rayner, J.M.V., 1979, A new approach to animal flight mechanics: Journal of Experimental Biology, v. 80, p. 17–54.

Rayner, J.M.V., 1988, The evolution of vertebrate flight: Biological Journal of the Linnean Society, v. 34, p. 269–287.

Ricqles, A.J., Padian, K., Horner, J.R., and Francillon-Viellot, H., 2000, Paleohistology of the bones of pterosaurs (Reptilia: Archosauria): anatomy, ontogeny, and biomechanical implications: Zoological Journal of the Linnean Society, v. 129, p. 349–385.

Roberts, T.J., Kram, R., Wey, P.G., and Taylor, C.R., 1998, Energetics of bipedal running, I. Metabolic cost of generating force: Journal of Experimental Biology, v. 201, p. 2745–2751.

Rohlf, F.J., 2001, tpsDig version 1.31, Program for digitizing of coordinates of landmarks and capture of outlines, available at: http://life.bio.sunysb.edu/morph/ (August 2003).

Romer, A.S., 1956, Osteology of the reptiles: Chicago, University of Chicago Press, 772 p.

Seeley, H.G., 1870, The Ornithosauria: an elementary study of the bones of pterodactyles, made from fossil remains found in the Cambridge Upper Greensand, and arranged in the Woodwardian Museum of the University of Cambridge: London, Dieghton Bell, 130 p.

Sereno, P.C., 1991, Basal archosaurs: phylogenetic relationships and functional implications: Journal of Vertebrate Paleontology, v. 11, suppl. 4, p. 1–51.

Sharov, A.G., 1971, Noviye lyetayushchie reptilii iz mezozoya Kazakhstana i Kirgizii: Trudy Paleontologicheskogo Instituta, Akademiia Nauk SSSR, v. 130, p. 104–113.

Sibley, D.A., 2001, The Sibley guide to bird life and behavior: New York, Alfred A. Knopf, 588 p.

Soemmerring, S.T. von, 1820, Über einen *Ornithocephalus brevirostris* der Vorwelt: Denkschriften der Königliche Akademic der Wissenschaften zu München, v. 3, p. 89–158.

Stanley, S.M., 1973, An explanation for Cope's Rule: Evolution, v. 27, p. 1–26.

Stein, R.S., 1975, Dynamic analysis of *Pteranodon longiceps*: a reptilian adaptation for flight: Journal of Paleontology, v. 49, p. 534–548.

Stepniewski, W.Z., and Keys, C.N., 1984, Rotary-wing aerodynamics: New York, Dover Publications, v. 1 and 2, 607 p.

Stokes, W.L., 1957, Pterodactyl tracks from the Morrison Formation: Journal of Paleontology, v. 31, p. 952–954.

Templin, R.J., 2000, The spectrum of animal flight: insects to pterosaurs: Progress in Aerospace Sciences, v. 36, p. 393–436.

Tennekes, H. 1996, The simple science of flight: Cambridge, Massachusetts Institute of Technology Press, 137 p.

Thulborn, T., 1990, Dinosaur tracks: London, Chapman and Hall, 410 p.

Tischlinger, H., and Frey, E., 2002, Ein *Rhamphorhynchus* (Pterosauria, Reptilia) mit ungewöhnlicher Flughauterhaltung aus dem Solnhofener Plattenkalk: Archaeopteryx: v. 20, p. 1–20.

Tucker, V.A., 1968, Respiratory exchange and evaporative water loss in a flying budgerigar: Journal of Experimental Biology, v. 48, p. 67–87.

Tucker, V., 1970, Energetic cost of locomotion in animals: Comparative Biochemistry and Physiology, v. 34, p. 841–846.

Tucker, V., 1973, Bird metabolism during flight: evaluation of a theory: Journal of Experimental Biology, v. 58, p. 689–709.

Tucker, V., 1975, The energetic cost of moving about: American Scientist, v. 63, p. 413–419.

Unwin, D.M., 1987, Pterosaur locomotion; joggers or waddlers?: Nature, v. 327, p. 13–14.

Unwin, D.M., 1988a, New remains of the pterosaur *Dimorphodon* (Pterosauria: Rhamphorhynchoidea) and the terrestrial ability of early pterosaurs: Modern Geology, v. 13, p. 57–68.

Unwin, D.M., 1988b, Extinction and survival in birds, *in* Larwood, G.P., ed., Extinction and survival in the fossil record: Oxford, Clarendon Press, p. 295–318.

Unwin, D.M., 1989, A predictive method for the identification of vertebrate ichnites and its application to pterosaur tracks, *in* Gillette, D.D., and Lockley, M.G., eds., Dinosaur tracks and traces: New York, Cambridge University Press, p. 259–274.

Unwin, D.M., 1995, Preliminary results of a phylogenetic analysis of the Pterosauria (Diapsida: Archosauria), *in* Sun, A., and Wang, Y., eds., Sixth symposium on Mesozoic terrestrial ecosystems and biota, short papers: Beijing, China, China Ocean Press, p. 69–72.

Unwin, D.M., and Bakhurina, N.N., 1994, *Sordes pilosus* and the nature of the pterosaur flight apparatus: Nature, v. 371, p. 62–64.

Unwin, D.M., Frey, E., Martill, D.M., Clarke, J.B., and Reiss, J., 1996, On the nature of the pteroid in pterosaurs: Proceedings of the Royal Society of London, v. B263, p. 45–52.

Unwin, D.M., and Henderson, D., 1999, Testing the terrestrial ability of pterosaurs with computer-based methods [abs.]: Journal of Vertebrate Paleontology, v. 19, p. 81A.

Unwin, D.M., and Lu, J., 1997, On *Zhejiangopterus* and the relationships of pterodactyloid pterosaurs: Historical Biology, v. 12, p. 199–210.

Vasquez, R.J., 1992, Functional osteology of the avian wrist and the evolution of the flapping flight: Journal of Morphology, v. 211, p. 259–268.

Vaughan, T.A., 1966, Morphology and flight characteristics of molossid bats: Journal of Mammalogy, v. 47, p. 249–260.

Viscain, F.V., and Farina, R.A. 1999, On the flight capabilities and the distribution of the giant bird *Argentavis magnificens* (Teratornithidae): Lethaia, v. 32, p. 271–278.

Walkden, S.L., 1925, Experimental study of "soaring" albatrosses: Nature, v. 116, p. 132–134.

Wellnhofer, P., 1970, Die Pterodactyloidea (Pterosauria) der Oberjura-Plattenkalke Suddeutschlands: Bayerische Akademie der Wissenschaften, Mathematisch-Wissenschaftlichen Klasse, Abhandlungen, v. 141, p. 1–133.

Wellnhofer, P., 1975, Die Rhamphorhynchoidea (Pterosauria) der Oberjura-Plattenkalke Suddeutschlands: Palaeontographica, v. A148, p. 1–33.

Wellnhofer, P., 1985, Neue Pterosaurier aus der Santana Formation (Aptian) of the Chapada do Aararipe, Brazil: Palaeontographica, v. A187, p. 105–182.

Wellnhofer, P., 1988, Terrestrial locomotion in pterosaurs: Historical Biology, v. 1, p. 3–16.

Wellnhofer, P., 1991a, The illustrated encyclopedia of pterosaurs: New York, Crescent Books, p. 192.

Wellnhofer, P., 1991b, Weitere Pterosaurierfunde aus der Santana Formation (Apt) der Chapada do Araripe, Brasilien: Palaeontographica, v. A215, p. 43–101.

West, G.B., Brown, J.H., and Enquist, B.J., 1999, The fourth dimension of life: fractal geometry and allometric scaling of organisms: Science, v. 284, p. 1677–1679.

Wild, R., 1978, Die flugsaurier (Reptilia, Pterosauria) aus der Oberen Trias von Cene bei Bergamo, Italien: Bolletino della Societa Paleontologica Italiana, v. 17, p. 176–257.

Wild, R., 1983, Über den Urspung der Flugsaurier, *in* Erwin Rutte-Festschrift: Weltenberger Ackademie, Kelheim, Weltenburg, p. 231–238.

Wild, R., 1984, A new pterosaur (Reptilia, Pterosauria) from the Upper Triassic (Norian) of Friuli, Italy: Gortania, Atti Museo Friulano di Storia Naturale, b. 5, p. 45–62.

Winkworth, S., 1985, *Pteranodon* flies again: New Scientist, v. 105, p. 32–33.

Witmer, L.M., Chatterjee, S., Franzosa, J., and Rowe, T., 2003, Neuroanatomy of flying reptiles: implications for the flight, head posture, and behavior of pterosaurs: Nature v. 425, p. 950–953.

Yalden, D.W., 1984, What was the Size of *Archaeopteryx*?: Zoological Journal of the Linnean Society, v. 82, p. 177–188.

Yalden, D.W., and Morris, P.P., 1975, The lives of bats: Newton, David and Charles, 247 p.

Young, C.C., 1973, Pterosaurian fauna from Wuerho, Sinkiang: Reports of Paleontological Expedition to Sinkiang (II): Kexue Chubanshe: Nanjing, China, p. 18–34.

Zittell, K.A. von, 1882, Über Flugsaurier aus dem lithographischen Schiefer Bayerns: Paläontographica, v. 29, p. 47–80.

Manuscript accepted by the Society September 4, 2003